W9-BID-557

THE LOST WORLD OF GENESIS ONE

Ancient Cosmology and the Origins Debate

JOHN H. WALTON

InterVarsity Press
P.O. Box 1400, Downers Grove, IL 60515-1426
World Wide Web: www.ivpress.com
E-mail: email@ivpress.com

InterVarsity Press® is the book-publishing division of InterVarsity Christian Fellowship/USA®, a movement of students and faculty active on campus at hundreds of universities, colleges and schools of nursing in the United States of America, and a member movement of the International Fellowship of Evangelical Students. For information about local and regional activities, write Public Relations Dept., InterVarsity Christian Fellowship/USA, 6400 Schroeder Rd., P.O. Box 7895, Madison, WI 53707-7895, or visit the IVCF website at <www.intervarsity.org>.

Design: Cindy Kiple
Images: lush landscape: David Ellison/iStockphoto

ISBN 978-0-8308-3704-5

Printed in the United States of America ∞

InterVarsity Press is committed to protecting the environment and to the responsible use of natural resources. As a member of Green Press Initiative we use recycled paper whenever possible. To learn more about the Green Press Initiative, visit <www.greenpressinitiative.org>.

Library of Congress Cataloging-in-Publication Data

Walton, John H., 1952-
The lost world of Genesis One: ancient cosmology and the origins debate / John H. Walton.
p. cm.
Includes bibliographical references (p.) and index.
ISBN 978-0-8308-3704-5 (pbk.: alk paper)
1. Bible. O.T. Genesis I—Criticism, interpretation, etc. 2. Biblical cosmology. 3. Creationism. 4. Cosmogony. I. Title.
BS651.W275 2009
231.7'65—dc22

2009011678

P *24 23 22 21 20 19 18 17 16 15 14 13 12 11 10 9 8 7 6 5 4 3*
Y *29 28 27 26 25 24 23 22 21 20 19 18 17 16 15 14 13 12 11 10 09*

Contents

Prologue

One of the principal attributes of God affirmed by Christians is that he is Creator. That conviction is foundational as we integrate our theology into our worldview. What all is entailed in viewing God as Creator? What does that affirmation imply for how we view ourselves and the world around us? These significant questions explain why discussions of theology and science so often intersect. Given the ways that both have developed in Western culture, especially in America, these questions also explain why the two often collide.

The first chapter of Genesis lies at the heart of our understanding of what the Bible communicates about God as Creator. Though simple in the majesty of its expression and the power of its scope, the chapter is anything but transparent. It is regrettable that an account of such beauty has become such a bloodied battleground, but that is indeed the case.

In this book I have proposed a reading of Genesis that I believe to be faithful to the context of the original audience and author, and one that preserves and enhances the theological vitality of this text. Along the way is opportunity to discuss numerous areas of controversy for Christians, including relating Genesis to modern science, especially evolution. Intelligent Design and creationism will be considered in light of the proposal, and I make some comments about the debate concerning public education.

The case is laid out in eighteen propositions, each presented succinctly and plainly so that those not trained in the technical fields involved can understand and use the information presented here. Whether the reader is an educated layperson who wants to know more, a pastor or youth pastor in a church, or a science teacher in public schools, he or she should find some stimulating ideas for thinking about the Bible, theology, faith and science.

Introduction

We like to think of the Bible possessively—*my* Bible, a rare heritage, a holy treasure, a spiritual heirloom. And well we should. The Bible is fresh and speaks to each of us as God's revelation of himself in a confusing world. It is ours and at times feels quite personal.

But we cannot afford to let this idea run away with us. The Old Testament *does* communicate to us and it was written for us, and for all humankind. But it was not written *to* us. It was written to Israel. It is God's revelation of himself to Israel and secondarily through Israel to everyone else. As obvious as this is, we must be aware of the implications of that simple statement. Since it was written to Israel, it is in a language that most of us do not understand, and therefore it requires translation. But the language is not the only aspect that needs to be translated. Language assumes a culture, operates in a culture, serves a culture, and is designed to communicate into the framework of a culture. Consequently, when we read a text written in another language and addressed to another culture, we must translate the culture as well as the language if we hope to understand the text fully.

As complicated as translating a foreign language can be, translating a foreign culture is infinitely more difficult. The problem lies in the act of translating. Translation involves lifting the ideas from their native context and relocating them in our own context. In some ways this is an imperialistic act and bound to create some distortion as we seek to organize information in the categories that are familiar to us. It is far too easy to let our own ideas creep in and subtly (or at times not so subtly) bend or twist the material to fit our own context.

On the level of words, for example, there are Hebrew words that simply do not have matching words in English. The Hebrew word *ḥesed* is a good example. The translators of the New American Standard Bible decided to adopt the combination word "lovingkindness" to render it. Other translations use a wide variety of words: loyalty, love, kindness and so on. The meaning of the word cannot easily be expressed in English, so using any word unavoidably distorts the text. English readers unaware of this could easily begin working from the English word and derive an interpretation of the text based on what that English word means to them, and thus risk bringing something to the text that was not there. Nevertheless translators have little choice but to take the word out of its linguistic context and try to squeeze it into ours—to clothe its meaning in English words that are inadequate to express the full meaning of the text.

When we move to the level of culture, the same type of problem occurs. The very act of trying to *translate* the culture requires taking it out of its context and fitting it into ours. What does the text mean when it describes Sarah as "beautiful"? One not only has to know the meaning of the word, but also must have some idea of what defines beauty in the ancient world. When the Bible speaks of something as elemental as marriage, we are not wrong to think of it as the establishment of a socially and legally recognized relationship between a man and a woman. But marriage carries a lot more social nuance than that in our culture and not necessarily similar at all to the social nuances in the ancient culture. When marriages are arranged and represent alliances between families and exchange of wealth, the institution fills a far different place in the culture than what we know when feelings of love predominate. In that light the word *marriage* means something vastly different in ancient culture, even though the word is translated properly. We would seriously distort the text and interpret it incorrectly if we imposed all of the aspects of marriage in our culture into the text and culture of the Bible. The minute anyone (professional or amateur) attempts to *translate* the culture, we run the risk of making the text communicate something it never intended.

Rather than translating the culture, then, we need to try to enter the culture. When people want to study the Bible seriously, one of the steps they take is to learn the language. As I teach language students, I am still always faced with the challenge of persuading them that they will not succeed simply by learning enough of the language to engage in translation. Truly learning the language requires leaving English behind, entering the world of the text and understanding the language in its Hebrew context without creating English words in their minds. They must understand the Hebrew as Hebrew text. This is the same with culture. We must make every attempt to set our English categories aside, to leave our cultural ideas behind, and try our best (as limited as the attempt might be) to understand the material in its cultural context without translating it.

How do we do this? How can we recover the way that an ancient culture thought and what categories and ideas and concepts were important to them? We have already noted that language is keyed to culture, and we may then also recognize that literature is a window to the culture that produced it. We can begin to understand the culture by becoming familiar with its literature. Undoubtedly this sounds like a circular argument: We can't interpret the literature without understanding the culture, and we can't understand the culture without interpreting the literature. If we were dealing only with the Bible, it would indeed be circular, because we have already adjusted it to our own cultural ways of thinking in our long familiarity with it. The key then is to be found in the literature from the rest of the ancient world. Here we will discover many insights into ancient categories, concepts and perspectives. Not only do we expect to find linkages, we do in fact find many such linkages that enhance our understanding of the Bible.

To compare the Old Testament to the literature of the ancient world is not to assume that we expect or find similarity at every point; but neither should we assume or expect differences at every point. We believe the nature of the Bible to be very different from anything else that was available in the ancient world. The very fact that we accept the Old Testament as God's revelation of himself distinguishes it from the literature of Mesopotamia or Egypt. For that matter, Egyptian literature was very different from Mesopotamian literature, and within Mesopotamia, Assyrian literature and Babylonian literature were far from homogeneous. To press the point further, Babylonian literature of the second millennium must be viewed as distinct from Babylonian literature of the first millennium. Finally we must recognize that in any given time period in any given culture in any given city, some people would have had different ideas than others. Having said all of this, we recognize at the same time that there is some common ground. Despite all the distinctions that existed across the ancient world, any given ancient culture was more similar to other ancient cultures than any of them are to Western American or European culture. Comparing the ancient cultures to one another will help us to see those common threads even as we become aware of the distinctions that separated them from one another. As we identify those common threads, we will begin to comprehend how the ancient world differed from our modern (or postmodern) world.

So to return to the illustration of marriage: we will understand the Israelite ideas of marriage much more accurately by becoming informed about marriage in Babylon or Egypt than we will by thinking of marriage in modern terms. Yet we will also find evidence to suggest that Babylonian customs and ideas were not always exactly like Israelite ones. The texts serve as sources of information for us to formulate the shape of each culture's ways of thinking. In most areas there is more similarity between Israel and its neighbors than there is between Israel and our twenty-first-century Western world. As another example, even though today we believe in one God, the God of Israel, and therefore share with them this basic element of faith, the views of deity in the ancient world served as the context for Israel's understanding of deity. It is true that the God of the Bible is far different from the gods of the ancient cultures. But Israel understood its God in reference to what others around them believed. As the Bible indicates, Israelites were continually drawn into the thinking of the cultures around them, whether they were adopting the gods and practices of those around them or whether they were struggling to see their God as distinct.

As a result, we are not looking at ancient literature to try to decide whether Israel borrowed from some of the literature that was known to them. It is to be expected that the Israelites held many concepts and perspectives in common with the rest of the ancient world. This is far different from suggesting literature was borrowed or copied. This is not even a case of Israel being influenced by the peoples around them. Rather we simply recognize the common conceptual worldview that existed in ancient times. We should therefore not speak of Israel being influenced by that world—they were part of that world.

To illustrate the idea, we must think of ways in which we are products of our own culture. For example, we do not borrow the idea of consumerism, nor are we influenced by it. We *are* consumers because we live in a capitalist society that is built on consumerism. We don't have to think about it or read about it. Even if we wanted to reject its principles we would find it difficult to identify all its different aspects and devise different ways of thinking. One could make similar observations about Aristotelian, Cartesian or Baconian forms of thought. We could speak of capitalism and the value of liberty. We could consider self-determinism and individualism. We could analyze our sense of personal rights and the nature of democracy. These are ideas and ways of thinking that make us who we are in the United States. Where did we learn the principles of naturalism or the nature of the universe? They are simply absorbed through the culture in which we live. One can find all of this in our literature, but we didn't learn it from our literature—it is simply part of our culture that we absorb, often with no alternatives even considered.

By recognizing the importance of the literatures of the ancient world for informing us about its cultures, we need not be concerned that the Bible must consequently be understood as just another piece of ancient mythology. We may well consider some of the literatures of Babylonia and Egypt as mythological, but that very mythology helps us to see the world as they saw it. The Canaanites or the Assyrians did not consider their myths to be made up works of the imagination. Mythology by its nature seeks to explain how the world works and how it came to work that way, and therefore includes a culture's "theory of origins." *We* sometimes label certain literature as "myth" because we do not believe that the world works that way. The label is a way of holding it at arm's length so as to clarify that we do not share that belief—particularly as it refers to involvement and activities of the gods. But for the people to whom that mythology belonged, it was a real description of deep beliefs. Their "mythology" expressed their beliefs concerning what made the world what it was; it expressed their theories of origins and of how their world worked.

By this definition, our modern mythology is represented by science—our own theories of origins and operations. Science provides what is generally viewed as the consensus concerning what the world is, how it works and how it came to be. Today, science makes no room for deity (though neither does it disprove deity), in contrast to the ancient explanations, which were filled with deity. For the Israelites, Genesis 1 offered explanations of their view of origins and operations, in the same way that mythologies served in the rest of the ancient world and that science serves our Western culture. It represents what the Israelites truly believed about how the world got to be how it is and how it works, though it is not presented as their own ideas, but as revelation from God. The fact that many people today share that biblical belief makes the term *mythology* unpalatable, but it should nevertheless be recognized that Genesis 1 serves the similar function of offering an explanation of origins and how the world operated, not only for Israel, but for people today who put their faith in the Bible.

PROPOSITION 1

Genesis 1 Is Ancient Cosmology

So what are the cultural ideas behind Genesis 1? Our first proposition is that Genesis 1 is ancient cosmology. That is, it does not attempt to describe cosmology in modern terms or address modern questions. The Israelites received no revelation to update or modify their "scientific" understanding of the cosmos. They did not know that stars were suns; they did not know that the earth was spherical and moving through space; they did not know that the sun was much further away than the moon, or even further than the birds flying in the air. They believed that the sky was material (not vaporous), solid enough to support the residence of deity as well as to hold back waters. In these ways, and many others, they thought about the cosmos in much the same way that anyone in the ancient world thought, and not at all like anyone thinks today.[1] And God did not think it important to revise their thinking.

Some Christians approach the text of Genesis as if it has modern science embedded in it or it dictates what modern science should look like. This approach to the text of Genesis 1 is called "concordism," as it seeks to give a modern scientific explanation for the details in the text. This represents one attempt to "translate" the culture and text for the modern reader. The problem is, we cannot translate their cosmology to our cosmology, nor should we. If we accept Genesis 1 as ancient cosmology, then we need to interpret it as ancient cosmology rather than translate it into modern cosmology. If we try to turn it into modern cosmology, we are making the text say something that it never said. It is not just a case of adding meaning (as more information has become available) it is a case of changing meaning. Since we view the text as authoritative, it is a dangerous thing to change the meaning of the text into something it never intended to say.

Another problem with concordism is that it assumes that the text should be understood in reference to current scientific consensus, which would mean that it would neither correspond to last century's scientific consensus nor to that which may develop in the next century. If God were intent on making his revelation correspond to science, we have to ask which science. We are well aware that science is dynamic rather than static. By its very nature science is in a constant state of flux. If we were to say that God's revelation corresponds to "true science" we adopt an idea contrary to the very nature of science. What is accepted as true today, may not be accepted as true tomorrow, because what science provides is the best explanation of the data at the time. This "best explanation" is accepted by consensus, and often with a few detractors. Science moves forward as ideas are tested and new ones replace old ones. So if God aligned revelation with one particular science, it would have been unintelligible to people who lived prior to the time of that science, and it would be obsolete to those who live after that time. We gain nothing by bringing God's revelation into accordance with today's science. In contrast, it makes perfect sense that God communicated his revelation to his immediate audience in terms they understood.

Since God did not deem it necessary to communicate a different way of imagining the world to Israel but was content for them to retain the native ancient cosmic geography, we can conclude that it was not God's purpose to reveal the details of cosmic geography (defined as the way one thinks about the shape of the cosmos). The shape of the earth, the nature of the sky, the locations of sun, moon and stars, are simply not of significance, and God could communicate what he desired regardless of one's cosmic geography. Concordism tries to figure out how there could have been waters above the sky (Gen 1:7), whereas the view proposed here maintains that this terminology is simply describing cosmic geography in Israelite terms to make a totally different point. (See the next proposition for details.)

If cosmic geography is culturally descriptive rather than revealed truth, it takes its place among many other biblical examples of culturally relative notions. For example, in the ancient world people believed that the seat of intelligence, emotion and personhood was in the internal organs, particularly the heart, but also the liver, kidneys and intestines. Many Bible translations use the English word "mind" when the Hebrew text refers to the entrails, showing the ways in which language and culture are interrelated. In modern language we still refer to the heart metaphorically as the seat of emotion. In the ancient world this was not metaphor, but physiology. Yet we must notice that when God wanted to talk to the Israelites about their intellect, emotions and will, he did not revise their ideas of physiology and feel compelled to reveal the function of the brain. Instead, he adopted the language of the culture to communicate in terms they understood. The idea that people think with their hearts describes physiology in ancient terms for the communication of other matters; it is not revelation concerning physiology. Consequently we need not try to come up with a physiology for our times that would explain how people think with their entrails. But a serious concordist would have to do so to save the reputation of the Bible. Concordists believe the Bible must agree—be in concord with—all the findings of contemporary science.

Through the entire Bible, there is not a single instance in which God revealed to Israel a science beyond their own culture. No passage offers a scientific perspective that was not common to the Old World science of antiquity.[2]

Beyond the issue of cosmic geography, there are a number of other cultural and potentially scientific issues to consider concerning how people thought in the ancient world. Several questions might be considered:

- What is the level and nature of God's involvement in the world?
- What is God's relationship to the cosmos? Is he manifested within the cosmos? Is he controlling it from outside?
- Is there such a thing as a "natural" world?
- What *is* the cosmos? A collection of material objects that operate on the basis of laws? A machine? A kingdom? A company? A residence?
- Is the account of creation the description of a manufacturing process or the communication of a concept?

These and many other questions will be addressed throughout this book. The answers proposed will not be determined by what best supports what we would prefer to think or by what will eliminate the most problems. Instead we strive to identify, truly and accurately, the thinking in the ancient world, the thinking in the world of the Bible, and to take that where it leads us, whether toward solutions or into more problems.

Before we begin moving through the remainder of the propositions that make up this book, one of the issues raised in the list above should be addressed immediately. That is, there is no concept of a "natural" world in ancient Near Eastern thinking. The dichotomy between natural and supernatural is a relatively recent one.

Deity pervaded the ancient world. Nothing happened independently of deity. The gods did not "intervene" because that would assume that there was a world of events outside of them that they could step into and out of. The Israelites, along with everyone else in the ancient world, believed instead that every event was the act of deity—that every plant that grew, every baby born, every drop of rain and every climatic disaster was an act of God. No "natural" laws governed the cosmos; deity ran the cosmos or was inherent in it. There were no "miracles" (in the sense of events deviating from that which was "natural"), there were only signs of the deity's activity (sometimes favorable, sometimes not). The idea that deity got things running then just stood back or engaged himself elsewhere (deism) would have been laughable in the ancient world because it was not even conceivable. As suggested by Richard Bube, if God were to unplug himself in that way from the cosmos, we and everything else in the cosmos would simply cease to exist.[3] There is nothing "natural" about the world in biblical theology, nor should there be in ours. This does not suggest that God micromanages the world,[4] only that he is thoroughly involved in the operations and functions of the world.

As a result, we should not expect anything in the Bible or in the rest of the ancient Near East to engage in the discussion of how God's level of creative activity relates to the "natural" world (i.e., what we call naturalistic process or the laws of nature). The categories of "natural" and "supernatural" have no meaning to them, let alone any interest (despite the fact that in our modern world such questions take center stage in the discussion). The ancients would never dream of addressing how things might have come into being without God or what "natural" processes he might have used. Notice that even the biblical text merges these perspectives when Genesis 1:24 says, "Let the earth bring forth living creatures" but then follows up with the conclusion in the very next verse, "So God made the animals."[5] All of these issues are modern issues imposed on the text and not the issues in the culture of the ancient world. We cannot expect the text to address them, nor can we configure the information of the text to force it to comply with the questions we long to have answered. We must take the text on its own terms—it is not written to us. Much to our dismay then, we will find that the text is impervious to many of the questions that consume us in today's dialogues. Though we long for the Bible to weigh in on these issues and give us biblical perspectives or answers, we dare not impose such an obligation on the text. God has chosen the agenda of the text, and we must be content with the wisdom of those choices. If we attempt to commandeer the text to address our issues, we distort it in the process.

As we begin our study of Genesis 1 then, we must be aware of the danger that lurks when we impose our own cultural ideas on the text without thinking. The Bible's message must not be subjected to cultural imperialism. Its message transcends the culture in which it originated, but the form in which the message was imbedded was fully permeated by the ancient culture. This was God's design and we ignore it at our peril. Sound interpretation proceeds from the belief that the divine and human authors were competent communicators and that we can therefore comprehend their communication. But to do so, we must respect the integrity of the author by refraining from replacing his message with our own. Though we cannot expect to be able to think like they thought, or read their minds, or penetrate very deeply into so much that is opaque to us in their culture, we can begin to see that there *are* other ways of thinking besides our own and begin to identify some of the ways in which we have been presumptuously ethnocentric. Though our understanding of ancient culture will always be limited, ancient literature is the key to a proper interpretation of the text, and sufficient amounts of it are available to allow us to make progress in our understanding.

Technical Support

These are sources where I have dealt with these issues in more depth:

"Ancient Near Eastern Background Studies." In *Dictionary for Theological Interpretation of the Bible,* edited by Kevin J. Vanhoozer et al., pp. 40-45. Grand Rapids: Baker Academic, 2005.

Ancient Near Eastern Thought and the Old Testament: Introducing the Conceptual World of the Hebrew Bible. Grand Rapids: Baker Academic, 2006.

Genesis. New International Version Application Commentary. Grand Rapids: Zondervan, 2001.

"Interpreting the Bible as an Ancient Near Eastern Document." In *Israel: Ancient Kingdom or Late Invention,* edited by Daniel I. Block, pp. 298-327. Nashville: Broadman & Holman, 2008.

PROPOSITION 2

Ancient Cosmology Is Function Oriented

WHAT DOES IT MEAN FOR SOMETHING to exist? It might seem like an odd question with perhaps an obvious answer, but it is not as simple as it may seem. For example, when we say that a chair exists, we are expressing a conclusion on the basis of an assumption that certain properties of the chair define it as existing. Without getting bogged down in philosophy, in our contemporary ways of thinking, a chair exists because it is material. We can detect it with our senses (particularly sight and touch). We can analyze what it is made from. These physical qualities are what make the chair real, and because of them we consider it to exist. But there are other ways to think about the question of existence.

For example, we might consider what we mean when we talk about a company "existing." It would clearly not be the same as a chair existing. Does a company exist when it has filed the appropriate papers of incorporation? Does it exist when it has a building or a website? In some sense the answer to these would have to be yes. But many would prefer to speak of a company as existing when it is doing business. Consider what is communicated when a small retail business frames and displays the first dollar bill from the first sale. As another alternative, consider a restaurant that is required to display its current permit from the city department of health. Without that permit, the restaurant could be said not to exist, for it cannot do any business. Here existence is connected to the authority that governs existence in relation to the function the business serves. It is the government permit that causes that restaurant to exist, and its existence is defined in functional terms.

The question of existence and the previous examples introduce a concept that philosophers refer to as "ontology." Most people do not use the word *ontology* on a regular basis, and so it can be confusing, but the concept it expresses is relatively simple. The ontology of *X* is what it means for *X* to exist. If we speak of the ontology of evil, we discuss what it means for evil to exist in the world. The ontology of a chair or a company would likewise ask what it means when we say they exist. How would we understand their existence? What is the principle quality of its existence? The view represented in our discussion of the chair would be labeled a "material ontology"—the belief that something exists by virtue of its physical properties and its ability to be experienced by the senses. The example of the company might be labeled a "functional ontology."

In a discussion of origins we need to focus on the ontology of the cosmos. What does it mean for the world or the cosmos (or the objects in it) to exist? How should we think about cosmic ontology? When we speak of cosmic ontology these days, it can be seen that our culture views existence, and therefore meaning, in material terms. Our material view of ontology in turn determines how we think about creation, and it is easy to see how. If ontology defines the terms of existence, and creation means to bring something into existence, then one's ontology sets the parameters by which one thinks about creation. Creation of a chair would be a very different process than the creation of a company. Since in our culture we believe that existence is material, we consequently believe that to create something means to bring its material properties into existence. Thus our discussions of origins tend to focus on material origins.

All of this probably sounds like a silly discussion to many people. Of course something exists because it has material properties; of course creation means to give something material properties! Many would be inclined to ask in their exasperation, what else could it be? But our example of a company above has already alerted us to another possibility. Is it possible to have a cosmic ontology that is function oriented and see creation (bringing something into existence) in those terms?

Even staying in the realm of English usage we can see that we don't always use the verb *create* in material terms. When we create a committee, create a curriculum, create havoc or create a masterpiece, we are not involved in a material manufacturing process. Though a curriculum, for instance, eventually takes a material form, the creation of the curriculum is more a process of organizing ideas and goals. To understand what it means to "create" a curriculum, we would have to decide what it means for a curriculum to exist. What would be the ontology of a curriculum? Whatever our answer might be, these examples should suggest that there are alternate ways of thinking about creative activity, even in our culture. If a curriculum's ontology is functional, then creating that curriculum involves function-giving activities.

With that background in mind, we need to return to the question of cosmic ontology. Most of us never consider alternative ontologies. Our culture has given us our beliefs about what it means for the cosmos to exist (material ontology; existence is material; creation is a material act) and many of us would not realize that these beliefs are the result of a choice. It is a testimony to the pervasive influence of culture that this material ontology seems so obvious as to prevent any thought that it is open to discussion.

As some of the above examples indicate, however, there are alternatives. If we are going to understand a creation account from the ancient world we must understand what they meant by "creation," and to do that we must consider their cosmic ontology instead of supplying our own. It is less important what we might think about ontology. If we are dealing with an ancient account we must ask questions about the world of that text: What did it mean to someone in the ancient world to say that the world existed? What sort of activity brought the world into that state of existence and meaning? What constituted a creative act?

In this book I propose that people in the ancient world believed that something existed not by virtue of its material properties, *but by virtue of its having a function in an ordered system.* Here I do not refer to an ordered system in scientific terms, but an ordered system in human terms, that is, in relation to society and culture. In this sort of functional ontology, the sun does not exist by virtue of its material properties, or even by its function as a burning ball of gas. Rather it exists by virtue of the role that it has in *its* sphere of existence, particularly in the way that it functions for humankind and human society. In theory, this way of thinking could result in something being included in the "existent" category in a material way, but still considered in the "nonexistent" category in functional terms (see the illustration of the restaurant mentioned above). In a functional ontology, to bring something into existence would require giving it a function or a role in an ordered system, rather than giving it material properties. Consequently, something could be manufactured physically but still not "exist" if it has not become functional.

Perhaps a modern example can help. If we think of "creating" a computer, we understand that there are many stages in the process. At the most basic level the casing and the electronics have to be manufactured, the keyboard and other peripherals designed and so forth. This is the basic production and manufacturing process—what we might call the material phase of production. After someone has assembled all those manufactured parts we might say that the computer exists. But another aspect involves writing the programs. Even after those programs are written, if the software has not been installed on the computer, its "existence" is meaningless—it cannot function. So there is a separate process of installing the software that makes the computer theoretically functional. But what if there is no power source (electric or battery)? This is another obstacle to the computer's existence. Adding a power source, we might now claim that its existence is finally and completely achieved. But what if no one sits at the keyboard or knows how to use or even desires to use it? It remains nonfunctional, and, for all intents and purposes, as if it did not exist. We can see that different observers might be inclined to attribute "existence" to the computer at different stages in the process.

In a functional ontology, all of the above steps are important in the definition of existence. Unless people (or gods) are there to benefit from functions, existence is not achieved. Unless something is integrated into a working, ordered system, it does not exist. Consequently, the actual creative act is to assign something its functioning role in the ordered system. That is what brings it into existence. Of course something must have physical properties before it can be given its function, but the critical question is, what stage is defined as "creation"?

In the ancient world they were not ignorant of the senses and the level at which objects could be perceived by the senses. They would have no difficulty understanding the physical nature of objects. The question here concerns not what they perceived but what they gave significance to. When we speak of a computer we are certainly aware of the tower casing, and it is obvious that someone manufactured that. But that fact does not occupy our attention, nor do we confuse the manufacturing of the tower casing with the "creation" of the computer. To say this in another way, our ontology focuses on what we believe to be most significant. In the ancient world, what was most crucial and significant to their understanding of existence was the way that the parts of the cosmos functioned, not their material status.

How can we know this? The evidence comes both from the biblical text and from the literature of the ancient world. The former is more important because, of course, it is possible for the biblical text to take a different view of ontology than the ancient world. Propositions 3-11 will be offering the biblical evidence. For now then, we can set the stage from the ancient Near Eastern literature. Then we will see in which ways the biblical perspective corresponds and in which ways it differs.

A number of ancient Near Eastern texts giving information about creation come from the Sumerians, the Babylonians and the Egyptians.[1] Full-fledged creation texts include the following:

Egyptian:

- Memphite Theology (featuring Ptah)
- Papyrus Leiden I 350 (Hermopolis, featuring Amun)
- Pyramid Texts, Coffin Texts and Book of the Dead (especially from Heliopolis, featuring Atum)

Babylonian:

- Atrahasis
- *Enuma Elish*

Other sorts of texts that are not in and of themselves creation texts but contain information about creation include the following:

Sumerian. Numerous Sumerian texts contain cosmogonic (cosmogony = an account of the origins of the cosmos) or cosmological statements. Myths make statements in passing and rituals at times contain mythological sections that are cosmogonic. Even genealogical lists of the gods are thought to give hints to the extent that cosmogony can be inferred from theogony (theogony = an account of the origins of deity). Narrative texts from Nippur (an early sacred center in southern Mesopotamia) give the god Enlil a prominent role, while texts from Eridu (considered by the Sumerians to be the first city in history) favor the god Enki. Prominent also are the disputation texts (e.g., Tree and Reed, and such texts which feature discussions between animals or plants) which often have cosmogonic introductions. Akkadian cosmological information is also found in incantation texts as well as in introductions to dedicatory inscriptions.[2]

Egyptian. The most important allusions are found in the wisdom text titled the Instruction of Merikare and in cosmological depictions such as that on the centograph of Seti I.

Additional creation material is found in the Hittite Kumarbi Cycle and perhaps in the Ugaritic Baal Cycle.

What we learn from these can be summarized under several headings:

- ***Shape of the cosmos.*** Old world cosmic geography is based on what they could observe from their vantage point, just as ours is based on what we are able to observe given our scientific information (including, e.g., math and physics). If water comes down, there must be some up there—so they all thought in terms of cosmic waters in the sky. If it doesn't come down all the time, something must hold the water back—so it was common to think of something somewhat solid (firmament). If there is something solid holding back the waters, something must hold up this firmament—so they thought of mountains or ropes or tent poles. Waters come up from the ground so there must be waters under the ground, yet something must hold the ground steady. On and on the logic goes, following fairly transparent paths. As with any cosmic geography, the theories about structures are developed to understand the functions and operations as they are experienced and observed. Creation texts described these structures being put into place so that the operations would commence or continue.

- ***Role of deity.*** In the transition from cosmic geography to the role of deity, it is important to note that in the Egyptian descriptions of cosmic geography, all of those elements that we might consider cosmic structures (firmament, sun, moon, air, earth, etc.) are depicted as gods. This is strong evidence that the Egyptians were more interested in the functions of these gods than in the actual material structures. The gods represented authority and jurisdiction. The attributes of the deities were manifested in the cosmic elements. The cosmos functioned as an extension of the gods, and the gods functioned within the cosmos. The Mesopotamian texts do not have the artistic depictions, but they confirm the same interests, as the gods are seen in close relationship to the elements of the cosmos. It is the divine decree or divine assignment that dictates the role and function of the various elements.
- ***Origins of cosmos and deity.*** With the functions of the cosmos and the jurisdiction of the deities so closely correlated, it is no surprise that we find the origins of the gods (theogony) connected to the origins of the cosmic elements (cosmogony). This coinciding of origins indicates that those origins are functional in nature.
- ***Divine conflict.*** *Theomachy* is a term that refers to battles among the gods. Particularly in the Babylonian creation epic, *Enuma*

Elish, creation is accomplished in the aftermath of a battle for control of the pantheon and the cosmos.

- ***Features.***

Nonfunctional. Nearly all the creation accounts of the ancient world start their story with no operational system in place. Egyptian texts talk about a singularity—nothing having yet been separated out. All is inert and undifferentiated. Similarly, one Sumerian text speaks of a time when there was darkness, no flow of water, nothing being produced, no rituals performed, and heaven and earth were still joined together. Even the gods were not yet there.[3] For an example in Egyptian literature, the god Atum is conceptualized as the primordial monad—the singularity embodying all the potential of the cosmos, from whom all things were separated and thereby were created.[4]

Primeval waters. Creation often begins with that which emerges from the waters—whether a deity or land (e.g., the Egyptian Primeval Hillock). These primeval waters are designated the "nonexistent" in Egyptian texts, a key indicator of their functional ontology. The god Atum is said to have developed "out of the Flood, out of the Waters, out of darkness, out of lostness."[5] The Waters is termed the "father of the gods."[6]

Naming. Names in the ancient world were associated with identity, role and function. Consequently, naming is a typical part of the creation narratives. The Egyptian Memphite Theology identifies the Creator as the one who pronounced the name of everything. *Enuma Elish* begins with neither the heavens and earth nor the gods having yet been named. In this it is clear that naming is a significant part of something's existence, and therefore of its creation.

Separating. This is the most common creative activity in Egyptian texts and is also observable in a number of Mesopotamian texts. Heavens and earth are most often separated. Even Hittite literature indicates this important step when one myth talks about cutting heaven and earth apart with a copper cutting tool.[7] Others include separation of the upper and lower waters and waters from land.

Creatures. It is interesting that living creatures are almost never included in the creation accounts. The only exception is in the Akkadian Disputation of Two Insects, which mentions classification by size and by wild or domesticated nature.

Human beings. Many accounts of creation include human beings. Texts speak of what they are made of (clay, blood of deity, breath of deity) but not in a chemical sense. These ingredients communicate instead the important issues of identity and relationship (see further in proposition 6).

Before we leave the ancient Near Eastern texts, a few specific texts should be noted. The Egyptian Papyrus Insinger is from the Ptolemaic period (dated to the second or third century B.C., though the manuscript is from about the first century A.D.). Toward the end of this piece of wisdom literature, the paragraph designated the twenty-fourth Instruction contains eighteen lines of what the creations describe as the hidden work of the god.

> He created light and darkness in which is every creature.
> He created the earth, begetting millions, swallowing them up and begetting again.
> He created day, month, and year through the commands of the lord of command.
> He created summer and winter through the rising and setting of Sothis.

He created food before those who are alive, the wonder of the fields.
He created the constellation of those that are in the sky, so that those on earth should learn them.
He created sweet water in it which all the lands desire.
He created the breath in the egg though there is no access to it.
He created birth in every womb from the semen which they receive.
He created sinews and bones out of the same semen.
He created going and coming in the whole earth through the trembling of the ground.
He created sleep to end weariness, waking for looking after food.
He created remedies to end illness, wine to end affliction.
He created the dream to show the way to the dreamer in his blindness.
He created life and death before him for the torment of the impious man.
He created wealth for truthfulness, poverty for falsehood.
He created work for the stupid man, food for the common man.
He created the succession of generations so as to make them live.[8]

Though this text dates from well into the Hellenistic period, the functional orientation is obvious. Another example selected from a millennium earlier (twelfth c. B.C.) and from the opposite end of the ancient world demonstrates how pervasive this perspective was. In the Babylonian creation epic, *Enuma Elish,* Marduk defeats the rebellious gods and then does his work of "creation" in tablet five, focusing on several key functional features:

- Lines 1-24 show Marduk organizing the celestial sphere: stars, constellations, the phases of the moon.

- Lines 25-45 are not represented in many of the translations included in the major anthologies of ancient texts. Even in their broken form, however, their basic content can be discerned.[9] In 38-40 Marduk makes the night and day and sets it up so that there is an equal amount of light hours and night hours over the course of the year.[10] On line 46 he fixes the watches of night and day. These creative activities have to do with organizing time.
- Lines 47-52 are more legible and deal with the creation of the clouds, wind, rain, and fog, and appointing himself to control them. Here the functions that concern the weather are created.
- Lines 53-58 tell of the harnessing of the waters of Tiamat for the purpose of providing the basis of agriculture. It includes the piling up of dirt, releasing the Tigris and Euphrates, and digging holes to manage the catchwater.
- Lines 59-68 conclude with the transition into the enthronement of Marduk and the building of his temple and the city of Babylon—the grand climax. It is no surprise that a creation text should ultimately be about the god who controls the cosmos and about the origin of his temple. We will see below that cosmic origins and temple origins are intricately intertwined.

Finally, in a Sumerian debate text still another millennium earlier (third millennium), The Debate Between Winter and Summer, Enlil is involved in creation in these same areas (day and night/time; fertility/food; sluices of heaven/weather and seasons):

> An [god's name] lifted his head in pride and brought forth a good day. He laid plans for ...... and spread the population wide. Enlil set his foot upon the earth like a great bull. En-

> lil, the king of all lands, set his mind to increasing the good day of abundance, to making the ...... night resplendent in celebration, to making flax grow, to making barley proliferate, to guaranteeing the spring floods at the quay, to making ...... lengthen (?) their days in abundance, to making Summer close the sluices of heaven, and to making Winter guarantee plentiful water at the quay.[11]

In conclusion, analysts of the ancient Near Eastern creation literature often observe that nothing material is actually made in these accounts. This is an intriguing observation. Scholars who have assumed that true acts of creation must by definition involve production of material objects are apparently baffled that all of these so-called creation texts have nothing of what these scholars would consider to be creation activities. I propose that the solution is to modify what we consider creation activities based on what we find in the literature. If we follow the sense of the literature and its ideas of creation, we find that people in the ancient Near East did not think of creation in terms of making material things—instead, everything is function oriented. The gods are beginning their own operations and are making all of the elements of the cosmos operational. Creation thus constituted bringing order to the cosmos from an originally nonfunctional condition. It is from this reading of the literature that we may deduce a functional ontology in the ancient world—that is, that they offer accounts of functional origins rather than accounts of material origins. Consequently, to create something (cause it to exist) in the ancient world means to give it a function, not material properties. We need to note the contrast: we tend to think of the cosmos as a machine and argue whether someone is running the machine or not. The ancient world viewed the cosmos more like a company or a kingdom.[12]

Would they have believed that their gods also manufactured the material? Absolutely, for nothing can be thought to stand apart from the gods. But they show little interest in material origins. Such issues were simply insignificant to them. If we paused to think about it, we might begin to wonder why material origins have taken on such central significance to us. Consider:

- As employees we pay little attention to the history of the company we work for. We are more interested in its corporate structure and what responsibilities each department has. We want to know about who reports to whom and who is in charge of certain operations and tasks.
- When we go to the theater, we may have passing interest in the construction of the set and stage works, but we understand that the play exists in the roles of the performers. When a person comes late and asks what has happened so far, the question is not answered by information about the costume designer, script writer and the hiring of the cast. Telling the person about all that would be offering the wrong sort of origins information.

Some sorts of origins are more important than other sorts of origins.

In summary, this chapter has noted that our own material definition of existence is only one of the possible ways to define existence. I have suggested that in the ancient world they defined it differently. They thought of existence as defined by having a function in an ordered system.

Technical Support

Clifford, Richard. *Creation Accounts in the Ancient Near East and the Bible.* Catholic Biblical Quarterly Monograph Series 26. Washington, D.C.: Catholic Biblical Association, 1994.

Hyers, Conrad. *The Meaning of Creation: Genesis and Modern Sci-*

ence. Atlanta: John Knox Press, 1984.
Simkins, Ronald A. *Creator and Creation: Nature in the Worldview of Ancient Israel.* Peabody, Mass.: Hendrickson, 1994.
Stek, John. "What Says the Scripture?" In *Portraits of Creation,* edited by H. J. van Till, pp. 203-65. Grand Rapids: Eerdmans, 1990.
Walton, John. "Creation." In *Dictionary of the Old Testament: Pentateuch,* edited by T. Desmond Alexander and David W. Baker, pp. 155-68. Downers Grove, Ill.: InterVarsity Press, 2003.

PROPOSITION 3

"Create" (Hebrew *bārāʾ*) Concerns Functions

THE PREVIOUS CHAPTER PRESENTED evidence that creation accounts in the ancient world characteristically showed interest in the functional level rather than the material level. Furthermore it proposed that the ancient world defined existence in terms of having a function in an ordered system. This functional ontology indicated that the line between existence and nonexistence was functional, not material.

We now turn our attention to the creation account in Genesis 1 to discover whether it will follow suit or not. Our first matter for discussion is the Hebrew verb *bārāʾ*, translated as "create" in verse 1. What exactly does it mean? Here we cannot be content with delving into the English verb "create"—though that shows an amazing amount of flexibility. Instead we must focus on the verb in Hebrew and how its users would have understood its meaning. If we are trying to understand whether the Israelites thought of existence in functional terms (like the rest of the ancient Near East) or material terms (like we tend to do), one of the places we might expect to find help is in observing what is involved in bringing something into existence. "Create" is the English word for bringing something into existence. If existence is defined in material terms, creating is a material activity. If existence is defined in functional terms, creating is a function-giving activity. We cannot assume that creating is a material activity just because our ontology happens to be material. We must let the word and its usage speak for itself.

It is interesting that many people who discuss Genesis 1 express an interest in interpreting the chapter "literally." By this they generally mean that it is to be taken exactly for what it says rather than to understand Genesis 1 simply in metaphoric, allegorical or symbolic terms. Of course we recognize that sometimes writers intend to communicate by means of metaphor or allegory. When someone insists that Genesis 1 should be interpreted literally it is often an expression of their conviction that the interpreter rather than the author has initiated another level of meaning. Our interpretive commitment is to read the text at what I will call "face value." I will have more to say about this in proposition 11. For the moment, let us consider the concept and challenge of "literal" interpretation.

The English reader must face a difficult fact: one cannot comprehend the literal meaning of a word in the Old Testament without knowing Hebrew or having access to the analysis by someone who does. It does us no good to know what "create" literally means—we have to know what *bārāʾ* literally means.[1] Before that leads to frustration or despair, we can recognize that even those without knowledge of Hebrew can check the data of the Hebrew analyst at some level. A quick review of words and how they work will help us all to see how this is so.

First, we recognize that there is no ancient dictionary of Hebrew that gives us the definitions of all of the words (especially not in English). Instead we rely on the careful work done by commentators and translators over the centuries. How do these scholars figure out the meaning of words? The same way all of us do in whatever language we speak—by usage.[2] The meanings of words are established and determined by the ways in which they are used. This includes the kinds of sentences they are used in, the words they can be compared to (synonyms or antonyms), and the words they are used in connection with. For nouns this means what verbs they take; for verbs it includes what subjects or objects are associated with them. It is context that tells us whether a word is used metaphorically or with an idiomatic or technical sense.[3] Consequently a scholar who says that a Hebrew word means this or that should offer evidence from usage to support his or her findings. Having been provided a list of references in such an analysis, even someone who does not know Hebrew can double check the data. So, for instance, when I say that all the occurrences of *bārāʾ* have God as the subject or implied subject, an English reader can look at all the occurrences and see that this is so.

Now the analysis can begin. What can be said about the Hebrew verb *bārāʾ?* First, there is no passage in the Old Testament that offers an explanatory gloss for *bārāʾ*—that is, that says "by *bārāʾ* I mean *X*." So, as usual, we must depend on circumstantial, contextual analysis: subjects, objects and related terms.

SUBJECTS

The verb *bārāʾ* occurs about fifty times in the Old Testament. As referred to above, deity is always either the subject or the implied subject (in passive constructions) of the verb. It can therefore be confidently asserted that the activity is inherently a divine activity and not one that humans can perform or participate in. This observation is widely discussed, and on this conclusion all commentators agree.

Objects

It is of interest that few commentators discuss the objects of the verb, but this is the most important issue for our analysis. Since we are exploring what constitutes creative activity (specifically, material or functional), then the nature of that which has been created is of utmost significance. If the objects of the verb are consistently material that would be important information; likewise if they are consistently functional. Of course the profile is unlikely to be so straightforward. Ambiguous contexts are bound to exist, so a bit of methodology must be discussed.

Theoretically, the verb could be broad enough to include either material or functional activity. For that matter, we might conclude that it involves (at least in some cases) both material and functional. Assuming that there will be ambiguous cases (and there are), it is important to see if we have any contexts which *must* be understood in material terms or which *must* be understood in functional terms. If all occurrences were either material or ambiguous, we could not claim support for a functional understanding. If all occurrences were either functional or ambiguous, we could not claim clear support for a material understanding. If there are clear examples that can be only functional, and other clear examples that can only be material, then we would conclude that the verb could work in either kind of context, and ambiguous cases would have to be dealt with on a case-by-case basis.

Table 1 provides a comprehensive list of the objects of *bārāʾ*.[4] (See p. 42.)

The grammatical objects of the verb can be summarized in the following categories:

cosmos (10, including new cosmos)
people in general (10)
specific groups of people (6)

Table 1

Reference	Object	Comments
Gen 1:1	heavens and earth	
Gen 1:21	creatures of the sea	
Gen 1:27	people	male and female
Gen 1:27 (2)	people	in his image
Gen 2:3	(none)	
Gen 2:4	heavens and earth	
Gen 5:1	people	likeness of God
Gen 5:2	people	male and female
Gen 5:2	people	
Gen 6:7	people	
Ex 34:10	wonders	parallel to *ʿāśâ* (made/did)
Num 16:30	something new (debatable)	earth swallowing rebels
Deut 4:32	people	
Ps 51:10	pure heart	
Ps 89:12	north and south	
Ps 89:47	people	for futility
Ps 102:18	people not yet created	to praise the Lord
Ps 104:30	creatures	renewing the face of the earth
Ps 148:5	celestial inhabitants	to praise the Lord
Eccles 12:1	you	
Is 4:5	cloud of smoke	
Is 40:26	starry host	called by name, kept track of
Is 40:28	ends of the earth	
Is 41:20	rivers flowing in desert	to meet needs of his people
Is 42:5	heavens	stretched out
Is 43:1	Jacob	= Israel
Is 43:7	everyone called by my name	for my glory
Is 43:15	Israel	
Is 45:7	darkness	parallel to forming light
Is 45:7	disaster	parallel to bringing prosperity
Is 45:8	heavens and earth	to produce salvation and righteousness
Is 45:12	people	
Is 45:18	earth	did not create it to be *(tōhû)*

specific individuals or types of individuals (5)
creatures (2)
phenomena (e.g., darkness) (10)
components of cosmic geography (3)
condition (1, pure heart)

This list shows that grammatical objects of the verb are not easily identified in material terms, and even when they are, it is questionable that the context is objectifying them.[5] That is, no clear example occurs that demands a material perspective for the verb, though many are ambiguous.[6] In contrast, a large percentage of the contexts require a functional understanding. These data cannot be used to prove a functional ontology, but they offer support that existence is viewed in functional rather than material terms, as is true throughout the rest of the ancient world. If the Israelites understood the word *bārāʾ* to convey creation in functional terms, then that is the most "literal" understanding that we can achieve. Such an understanding does not represent an attempt to accommodate modern science or to neutralize the biblical text. The truest meaning of a text is found in what the author and hearers would have thought.

This view finds support from an unexpected direction. It has long been observed that in the contexts of *bārāʾ* no materials for the creative act are ever mentioned, and an investigation of all the passages mentioned above substantiate that claim. How interesting it is that these scholars then draw the conclusion that *bārāʾ* implies creation out of nothing *(ex nihilo)*. One can see with a moment of thought that such a conclusion assumes that "create" is a material activity. To expand their reasoning for clarity's sake here: Since "create" is a material activity (assumed on their part), and since the contexts never mention the materials used (as demonstrated by the evidence), then the material object must have been brought into existence without using other materials (i.e., out of nothing). But one can see that the whole line of reasoning only works if one can assume that *bārāʾ* is a material activity. In contrast, if, as the analysis of objects presented above suggests, *bārāʾ* is a functional activity, it would be ludicrous to expect that materials are being used in the activity. In other words, the absence of reference to materials, rather than suggesting material creation out of nothing, is better explained as indication that *bārāʾ* is not a material activity but a functional one. This is not a view that has been rejected by other scholars; it is simply one they have never considered because their material ontology was a blind presupposition for which no alternative was ever considered.

An important caveat must be noted at this point. If we conclude that Genesis 1 is not an account of material origins, we are not thereby suggesting that God is not responsible for material origins. I firmly believe that God *is* fully responsible for material origins, and that, in fact, material origins do involve at some point creation out of nothing. But that theological question is not the one we are asking. We are asking a textual question: What sort of origins account do we find in Genesis 1? Or what aspect of origins is addressed in Genesis 1? Most interpreters have generally thought that Genesis 1 contains an account of material origins because that was the only sort of origins that our material culture was interested in. It wasn't that scholars examined all the possible levels at which origins could be discussed; they presupposed the material aspect.

Finally, we must put the verb *bārāʾ* in its context in verse 1 where it tells us that "in the beginning God created the heavens and the earth." One immediate question that would occur is, beginning of what? The answer is not transparent. We must ask what "beginning" refers to and how verse 1 functions in relation to the rest of the context.[7]

Beginning

In Hebrew usage this adverb typically introduces a period of time rather than a point in time.[8] We can most easily see this in Job 8:7, which speaks of the early part of Job's life, and Jeremiah 28:1, which refers to the beginning period of Zedekiah's reign. This usage happens to correspond with ideas that are reflected in ancient Near Eastern creation texts. Egyptian texts refer to the "first occasion," which implies the first occurrence of an event that is to be repeated or continued. In Akkadian the comparable term to the Hebrew refers to the first part or first installment. All of this information leads us to conclude that the "beginning" is a way of talking about the seven-day *period* rather than a *point* in time prior to the seven days.

The Role of Verse 1

If the "beginning" refers to the seven-day period rather than to a point in time before the seven-day period, then we would conclude that the first verse does not record a separate act of creation that occurred prior to the seven days—but that in fact the creation that it refers to is recounted in the seven days. This suggests that verse 1 serves as a literary introduction to the rest of the chapter. This suggestion is confirmed by the fact that Genesis 2:1 concludes the seven-day report with the statement that the "heavens and earth were completed," indicating that the creation of the heavens and earth was the work of the seven days, not something that preceded them.

Such a conclusion is also supported by the overall structure of the book of Genesis. All commentators have recognized the recurrent transitionary formula "This is the account *(tôlĕdôt)* of . . ." used eleven times by the author to identify the sections of the book of Genesis. This shows us that the author of Genesis indeed did use initial statements as literary introductions to sections. The first of these occurs in Genesis 2:4 as the first transition from the seven-day cosmogony to the Garden of Eden account. As a transitionary phrase it links what has come before to what comes next. Sometimes what follows is genealogical information that offers information about, for example, what became of Esau or Ishmael. Other times it is followed by narratives that offer information concerning, for instance, what came of Terah's family (thus the stories of Abram). The point is that this formula can only continue an already established sequence—it cannot begin that sequence.

The word "beginning" would be the logical term to introduce such a sequence. It would indicate the initial period, while the *tôlĕdôt* sections would introduce successive periods. If this were the case, the book would now have twelve formally designated sections (much more logical than eleven, considering the numbers that have symbolic significance in the Bible).

The proposals of this chapter can be summarized by the following expanded interpretive translation of verse 1: "In the initial period, God created by assigning functions throughout the heavens and the earth, and this is how he did it." The chapter *does* involve creative activities, but all in relation to the way that the ancient world thought about creation and existence: by naming, separating and assigning functions and roles in an ordered system. This was accomplished in the seven-day period that the text calls "the beginning." Genesis 2:3 comes back to this in its summary as it indicates the completion of the *bārāʾ* activities over the seven-day period.

TECHNICAL SUPPORT

Stek, John. "What Says the Scripture?" In *Portraits of Creation: Biblical and Scientific Perspectives on the World's Formation*, edited by H. J. van Till, pp. 203-65. Grand Rapids: Eerdmans, 1990.

PROPOSITION 4

The Beginning State in Genesis 1 Is Nonfunctional

If existence in the ancient world was best defined in functional terms rather than material ones, as suggested in previous chapters, and "create" is the activity that brings the transition from nonexistence to existence, then "creation" would also be a functional activity (as suggested for the Hebrew terminology in chapter 3). Further evidence should then be found in how creation accounts describe the "before" and "after" conditions. If the text offered an account of material origins, we would expect it to begin with no material. If the text offered an account of functional origins, we would expect it to begin with no functions.

Genesis 1 offers its starting point in verse 2, where it describes the earth as *tōhû* and *bōhû*. These terms are translated in a variety of ways in the most well-known English translations but with little true variation:

KJV, NASV: Formless and void
ESV, NKJV: Without form and void
NIV, NLT: Formless and empty
NRSV: A formless void
NJPS: Unformed and void
Net Bible: Without shape and empty

NCV: Empty and had no form

In contrast, detailed technical studies on the terms point in other directions. For example, David Tsumura, after a full semantic analysis, translates *tōhû* as "unproductive" rather than descriptive of something without physical form or shape.[1] As with our previous word study in chapter three, we must again take a look at the usage of the term to understand its meaning. In this study we must focus our attention on *tōhû* because the second term, *bōhû*, occurs only three times, and in all three is used in combination with *tōhû*. The Hebrew word *tōhû* occurs twenty times, as follows:

Table 2

Deut 32:10	parallel to the wilderness; described by "howling"
1 Sam 12:21	descriptive of idols who can accomplish nothing
Job 6:18	wasteland away from wadis where caravans perish for lack of water
Job 12:24	wandering in a trackless waste
Job 26:7	what the north is stretched over
Psalm 107:40	wandering in a trackless waste
Is 24:10	a *tōhû* settlement is described as desolate
Is 29:21	with *tōhû* they turn aside righteousness (similar to Is 59:4)
Is 34:11	measuring line of *tōhû* and plumb stone of *bōhû*
Is 40:17	worthlessness of the nations; parallel to "nothingness" and the "end"(?)
Is 40:23	rulers of the world made as *tōhû;* parallel to "nothingness"
Is 41:29	images are wind and *tōhû;* parallel to "end"(?) of their deeds
Is 44:9	all who make images are *tōhû;* parallel to without profit
Is 45:18	God did not bring it into existence *tōhû;* but in contrast formed it for habitation (intended function)
Is 45:19	Israelites not instructed to seek God in waste places; parallel to land of darkness
Is 49:4	expending one's strength to no purpose *(tōhû)*
Is 59:4	describes relying on empty arguments or worthless words (i.e., dissembling); parallel to that which is false or worthless
Jer 4:23	description of *tōhû* and *bōhû:* light gone, mountains quaking, no people, no birds, fruitful lands waste, towns in ruins

Studying this list, one can see nothing in these contexts that would lead us to believe that *tōhû* has anything to do with material form. The contexts in which they occur and the words and phrases used in parallel suggest rather that the word describes that which is nonfunctional, having no purpose and generally unproductive in human terms. Applying it as a descriptive term to nouns that represent geographical areas, nations, cities, people or idols all suggest the same conclusion. A word that had to do with material shape would not serve well in these contexts.

Why then has the term been so consistently translated as a reference to the absence of material form? One can only surmise that the translation tradition has been driven by the predominant material focus of the cultures that produced the translations. We must never forget that translation is the most basic act of interpretation. One cannot convey words meaningfully from a source language to a target language without first determining what they think the text means to say. If the translators were interpreting the text as an account of material origins, it is no surprise that *tōhû* was translated in material terms.

But even the material translation of *tōhû* could not obscure what is clear in verse 2: here at the beginning of the creation process, there is already material in existence—the waters of the deep. These primeval cosmic waters are the classic form that nonexistence takes in the functionally oriented ancient world.

Given the semantic information presented above and the treatment in the technical literature, we propose that *tōhû* and *bōhû* together convey the idea of nonexistence (in their functional ontology), that is, that the earth is described as not yet functioning in an ordered system. (Functional) creation has not yet taken place and therefore there is only (functional) nonexistence.

With this concept in mind, we return to Job 26:7: "He spreads out the northern (skies) over empty space *(tōhû);* he suspends the earth over nothing." The word translated "nothing" occurs only here in the Old Testament but is very important as it is parallel to *tōhû* in the passage. Technical analysis leads me to the conclusion that Job 26:7 describes the creation of heaven and earth in relation to the "nonexistent" cosmic waters above and below.[2] This provides further evidence that *tōhû* refers to the functionally nonexistent, which it represents geographically in the cosmic waters and the deserts as is common in the ancient Near Eastern texts. Thus the adjective *tōhû* could be used to refer

- to the precosmic condition (the beginning state in Genesis);
- to the functionless cosmic waters;
- or in the ordered creation to those places on which order had not been imposed, the desert and the cosmic waters above and below—surrounding the ordered cosmos.

The creation account in Genesis 1 can then be seen to begin with no functions rather than with no material. At this point, however, it is important to establish what we mean when we talk of functions. In our culture we even think of functions in material terms. We describe functions in scientific terms and understand function as a result of material properties. So we might describe the sun functionally as a burning ball of gas that projects heat and light, and which, by virtue of its gravitational pull, holds the solar system in orbit around it. In contrast, in the ancient world, function was not the result of material properties, but the result of purpose. The sun looks down on all and is associated with the god of justice. It functions as a marker for time and seasons. When the ancient texts talk about how something functions in an ordered system, the system under discussion is not a cosmic or ecological system. It is a system inhabited by beings. In the ancient Near East the functions were focused on the gods, who had created everything to work for their benefit and under their authority.

In the Old Testament God has no needs and focuses functionality around people. We will see increasing evidence of this understanding as we move through the remainder of Genesis 1. Consequently, functionality cannot exist without people in the picture. In Genesis people are not put in place until day six, but functionality is established with their needs and situation in mind.

This conclusion is further supported by the meaning of the repeated formula "it was good," which I propose refers to "functioning properly." Such a conclusion is not arbitrary but based on the context. Throughout Genesis 1 any number of possible meanings have been proposed for "good." In the history of interpretation it has often been understood in moral/ethical terms or as a reference to the quality of the workmanship. While the Hebrew term could be used in any of those ways, the context indicates a different direction. We can find out what the author means when saying all of these things are "good" by inquiring what it would mean for something *not* to be good. Fortunately the near context offers us just such an opportunity: "It is not good for the man to be alone" (Gen 2:18). This verse has nothing to do with moral perfection or quality of workmanship—it is a comment concerning function. The human condition is not functionally complete without the woman. Thus throughout Genesis 1 the refrain "it was good" expressed the functional readiness of the cosmos for human beings. Readers were assured that all functions were operating well and in accord with God's purposes and direction. Moreover the order and function established and maintained by God renders the cosmos both purposeful and intelligible. So there is reason or motivation for studying the detailed nature of creation, which we now call science, even if the ancient Hebrews didn't take up this particular study.

Based on the above assessment of the beginning state as it is presented in Genesis, we are now in a position to compare it to what we find in the ancient world. In the ancient Near East the precosmic condition is neither an abstraction ("Chaos") nor a personified adversary. But the primordial sea, which is the principal element of the precreation condition, *is* personified by Nammu in Sumer and by Nun in Egypt, and it can be perceived in an adversarial role.

More specifically, Egyptian texts describe the precosmic condition both in terms of what is lacking as well as by its positive features. That which is absent includes the spatial world (not yet separated), inhabitable places, life/death, procreation, time, conflict and diversity.[3] Positive features include limitless waters and total darkness.[4] Everything is brought into existence by being differentiated. The "after" picture is consequently one of inestimable diversity.[5]

When Sumerian and Akkadian sources document creation activities, we can observe both the situation before and after the activity, as well as what sorts of verbs are used. All of this helps to determine the focus of the creative activity. Many examples exist, but here I will present just one as an illustration, a few lines from the Sumerian text NBC[6] 11108:

> Earth was in darkness, the lower world was [invi]sible;
> The waters did not flow through the opening (in the earth),
> Nothing was produced, on the vast earth the furrow had not been made.
> The high priest of Enlil did not exist,
> The rites of purification were not carried out,
> The h[ierodul]e(?) of heaven was not adorned, she did not proclaim [the praises?]
> Heaven and earth were joined to each other (forming) a unit, they were not [married].[7]

The "before" picture here is composed both of what *is* present—darkness, water and the nondiscrete heaven and earth—and of what is *not:* the absence of productivity, of the gods and of the operation of the cult. Creative activities then alter this landscape. All of this indicates that cosmic creation in the ancient world was not viewed primarily as a process by which matter was brought into being, but as a process by which functions, roles, order, jurisdiction, organization and stability were established. This defines creation in the ancient world and in turn demonstrates that ontology was focused on something's functional status rather than its material status.

In summary, the evidence in this chapter from the Old Testament as well as from the ancient Near East suggests that both defined the pre-creation state in similar terms and as featuring an absence of functions rather than an absence of material. Such information supports the idea that their concept of existence was linked to functionality and that creation was an activity of bringing functionality to a nonfunctional condition rather than bringing material substance to a situation in which matter was absent. The evidence of matter (the waters of the deep in Gen 1:2) in the precreation state then supports this view.

Technical Support

Tsumura, David. *Creation and Destruction.* Winona Lake, Ind.: Eisenbrauns, 2005.

PROPOSITION 5

Days One to Three in Genesis 1 Establish Functions

Day One

Why didn't God simply call light "light"? This was one of the questions that first got me started on the journey that has resulted in the interpretation of Genesis 1 presented in this book. It was not the function orientation found in the ancient Near Eastern literature that changed my way of thinking about Genesis 1—it was the text of Genesis 1. The whole process begins with verse 5, the concluding verse of the account of day one:

> God called the light "day" and the darkness he called "night." And there was evening and there was morning—the first day. (NIV)

First of all it should be observed that light is never treated as a material object in the ancient Near East, despite our modern physics. It is rather thought of as a condition, just as darkness is. So even if light were being created, one would not be able to make the claim that this is a material act. In fact, however, light itself is not the focus of this day's activities. What is the text talking about when it indicates that God called the light "day"? After all, that is not what light is. The solution is not difficult to find. Some would even consider it transparent and hardly worth even noticing. If something connected with light is named "day" we can deduce that it is not light itself, but the period of light, for that is what "day" *is.* Since "day" is a period of light, and "day" is the name given, we conclude that we are dealing with a rhetorical device called metonymy in which a noun can reasonably be extended to a related concept.[1] In this case then, the author intends for us to understand the word "light" to mean a period of light. Otherwise the verse would not make sense. As a result, "God called the period of light 'day' and the period of darkness he called 'night.'"[2]

With this information from verse 5, we can now proceed backward through the text to verse 4. There we are told that "God separated the light from the darkness." Again we note that this statement does not make any sense if light and/or darkness are viewed as material objects. They cannot logically be separated, because by definition they cannot exist together in any meaningful scientific or material way. The solution of verse 5 works equally well here as the verse takes on its obvious meaning with God separating the period of light from the period of darkness. These are the distinct periods that are then named day and night in verse 5. So far so good.

Now comes the clincher. If "light" refers to a period of light in verse 5 and in verse 4, consistency demands that we extend the same understanding to verse 3, and here is where the "aha!" moment occurs. We are compelled by the demands of verses 4 and 5 to translate verse 3 as "God said, 'Let there be a period of light.'" If we had previously been inclined to treat this as an act of material creation, we can no longer sustain that opinion. For since what is called into existence is a period of light that is distinguished from a period of darkness and that is named "day," we must inevitably consider day one as describing the creation of time. The basis for time is the invariable alteration between periods of light and periods of darkness. This *is* a creative act, but it is creation in a *functional* sense, not a *material* one.

This interpretation solves the long-standing conundrum of why evening is named before morning. There had been darkness in the precreation condition. When God called forth a period of light and distinguished it from this period of darkness, the "time" system that was set up required transitions between these two established periods. Since the period of light had been called forth, the first transition was evening (into the period of darkness) and the second was morning (into the period of light). Thus the great cycle of time was put in place by the Creator. As his first act he mixed time into the features of the cosmos that would serve the needs of the human beings he was going to place in its midst.

A second conundrum that this resolves is the detail that many have found baffling over the ages as they ask, How could there be light on day one when the sun is not created until day four? Two observations can now be made: First, this is less of a problem when we are dealing with "time" in day one rather than specifically with "light." But this does not really resolve the problem without the second observation: If creation is understood in functional terms, the order of events concerns functional issues, not material ones. Time is much more important than the sun—in fact, the sun is not a function, it only *has* functions. It is a mere functionary. More about this in the next chapter.

DAY TWO

Day two has been problematic at a number of different levels. In antiquity people routinely believed that the sky was solid.[3] As history progressed through the periods of scholasticism, the Renaissance, the Copernican revolution and the Enlightenment, verse 2 became more difficult to handle. For if the Hebrew term is to be taken in its normal contextual sense, it indicates that God made a solid dome to hold up waters above the earth. The choice of saying the Bible was wrong was deemed unacceptable, but the idea of rendering the word in a way that could tolerate modern scientific thinking could not be considered preferable in that it manipulated the text to say something that it had never said. We cannot think that we can interpret the word "expanse/firmament" as simply the sky or the atmosphere if that is not what the author meant by it when he used it and not what the audience would have understood by the word. As we discussed in the first chapter, we cannot force Genesis to speak to some later science.

We may find some escape from the problem, however, as we continue to think about creation as ultimately concerned with the functional rather than the material. If this is not an account of material origins, then Genesis 1 is affirming nothing about the material world. Whether or not there actually are cosmic waters being held back by a solid dome does not matter. That material cosmic geography is simply what was familiar to them and was used to communicate something that is functional in nature. Instead of objectifying this water barrier, we should focus on the important twofold cosmic function it played. Its first role was to create the space in which people could live. The second and more significant function was to serve as a mechanism by which precipitation was controlled—the means by which weather operated. Order in the cosmos (for people especially) depended on the right amount of precipitation. Too little and we starve; too much and we are overwhelmed. The cosmic waters posed a continual threat, and the "firmament" had been created as a means of establishing cosmic order. That we do not retain the cosmic geography of the ancient world that featured a solid barrier holding back waters does not change the fact that our understanding of the Creator includes his role in setting up and maintaining a weather system. The material terms used in day two reflect accommodation to the way the ancient audience thought about the world. But it doesn't matter what one's material cosmic geography might look like—primitive or sophisticated—the point remains that on the second day, God established the functions that serve as the basis for weather.

DAY THREE

It is amazing to notice at this point that some interpreters are troubled by their observation that God doesn't make anything on day three. We can imagine their quandary—how can this be included in a creation account if God doesn't make anything on this day? By this point in the book, the reader can see the solution easily. Day three is only a problem if this is an account of material origins. If it is understood as an account of functional origins, there is no need for God to make something. Instead, we ask what function(s) were set up, and to that question we find ready answers.

First of all we note that just as day two separated and differentiated cosmic space, so day three differentiates terrestrial space. The act of separating, a key creation activity from a functional perspective, continues in prominence. Commonly in the ancient literature, these same differentiations can be seen.

Even as some commentators ponder the absence of material creation in day three, others often observe that the day seems to contain two separate acts (water/dry land and vegetation). From a functional perspective, the soil, the water and the principle of seed bearing are all very much related as essential to the production of food. The emergence of dry land from the waters is a common element in Egyptian cosmology, and there it has a definite referent. That is, the emergence of the primeval hillock in cosmology reflects the yearly reality of the fertile soil emerging in the aftermath of the inundation of the Nile. Thus it is clear that the emergence of dry land is associated with the growing of food.

Day three reflects the wonder of the ancient world at the whole idea that plants grew, dropped seed, and that more of the same plant came from that tiny seed. The cycle of vegetation, the principles of fertilization, the blessing of fecundity—all of these were seen as part of the amazing provision of food so necessary for people to survive.

So on day one God created the basis for time; day two the basis for weather; and day three the basis for food. These three great functions—time, weather and food—are the foundation of life. If we desire to see the greatest work of the Creator, it is not to be found in the materials that he brought together—it is that he brought them together in such a way that they work. Perhaps we can feel the same wonder when we consider how, even given all that we know about the physiology of the eye, that beyond all of our material understanding, through these bundles of tissue we can see. We should never lose the wonder of this. Functions are far more important than materials.

We should not be surprised to find that the three major functions introduced in the first three days of Genesis 1 are also prominent in ancient Near Eastern texts. These texts have already been cited in chapter two. Note again the three lines near the beginning of Papyrus Insinger:

> He created day, month, and year through the commands of the lord of command.
> He created summer and winter through the rising and setting of Sothis.
> He created food before those who are alive, the wonder of the fields.[4]

Likewise in Marduk's creative activity in *Enuma Elish* tablet five:

- Lines 38-40: night and day
- Lines 47-52: creation of the clouds, wind, rain and fog
- Lines 53-58: harnessing of the waters of Tiamat for the purpose of providing the basis of agriculture, piling up of dirt, releasing the Tigris and Euphrates, and digging holes to manage the catchwater[5]

But these functions feature prominently not just in other ancient cosmologies. In Genesis, after the cosmos is ordered, a crisis leads God to return the cosmos to an unordered, nonfunctional state by means of a flood. Here the cosmic waters are let loose from their boundaries and again the earth becomes nonfunctional. What follows is a re-creation text as the land emerges again from the waters and the blessing is reiterated.[6] Of greatest interest, in that context God makes the Creator's promise in Genesis 8:22:

> As long as the earth endures,
> Seedtime and harvest,
> Cold and heat,
> Summer and winter,
> Day and night
> Will never cease.

Here we find the same three major functions in reverse order: food, weather and time, never to cease. The author is well aware that these are the main categories in the operation of this world that God has organized.

In this chapter we have attempted to establish, first, that functional concerns rather than material ones dominate the account. Indeed the only appearance of what might be considered material in these three days is the firmament—the very thing that we are inclined to dismiss as not part of the material cosmos as we understand it. In contrast the functions of time, weather and food can be clearly seen in the text and recognized as significant in ancient Near Eastern cosmologies. More importantly, we can see that the prominence of these three functions is common to the ancient world. Perspectives on the material universe will vary from era to era and culture to culture. It would be no surprise then that God's creative work should be proclaimed relative to those issues that serve as the universal foundation of how people encounter the cosmos.

We should not worry about the question of "truth" with regard to the Bible's use of Old World science. As we mentioned before, some scientific framework needs to be adopted, and all scientific frameworks are dynamic and subject to change. Adoption of the framework of the target audience is most logical. The Old World science found in the Bible would not be considered "wrong" or "false" as much as it would just offer a perspective from a different vantage point. Even today we can consider it true that the sky is blue, that the sun sets and that the moon shines. But we know that these are scientifically misleading statements. Science, however, simply offers one way of viewing the world, and it does not have a corner on truth. The Old World science in the Bible offers the perspective of the earthbound observer. One could contend that there are some ways in which it is more true that the earth is the center of the cosmos. This does not mean to suggest that there are many truths, but that there are many possible different perspectives that can each offer truthful information. The way any culture describes the makeup of the material cosmos may vary considerably from how another might. A century ago the idea of an expanding universe would have seemed ludicrous, while today the steady-state universe has fallen into disfavor. This is all part of fine-tuning cosmic geography.

God did not give Israel a revised cosmic geography—he revealed his Creator role through the cosmic geography that they had, because the shape of the material world did not matter. His creative work focused on functions, and therefore he communicated that he was the one who set up the functions and who keeps the operations going, regardless of how we envision the material shape. This creation account did not concern the material shape of the cosmos, but rather its functions.

TECHNICAL SUPPORT

Seely, P. "The Firmament and the Water Above." *Westminster Theological Journal* 54 (1992): 31-46.

———. "The Geographical Meaning of 'Earth' and 'Seas' in Genesis 1:10." *Westminster Theological Journal* 59 (1997): 231-55.

PROPOSITION 6

Days Four to Six in Genesis 1 Install Functionaries

In the account of days four through six we see a shift in the focus. While a functional orientation is still obvious, God is not setting up functions as much as he is installing functionaries. In some cases the functionaries will be involved in carrying out the functions (especially the role of the celestial bodies in marking periods of time), but in most cases the functionaries simply carry out their own functions in the spheres delineated in the first three days (time, cosmic space, terrestrial space). The assignment of functionaries to their tasks and realms is equally an act of creation. Days four through six are literarily parallel to days one through three, as has long been recognized, but the literary structure is secondary (see chapter 13).

Day Four

In the report of this day the functional orientation can be clearly seen. The text offers no indication of the material nature of the celestial bodies, and all that it says of their material placement is that they are in the firmament/expanse. This is, of course, problematic if one is trying to understand the text scientifically. On the functional side of the equation, we find that they separate day and night (thus the link to day one), that they provide light and that they serve for "signs, seasons, days and years." Finally we are told that their function is to govern the day and night—the closest the text comes to personification.

Again we point out that these are not scientific functions but human-oriented functions. In this regard it should be noted that the fourfold description of functions (signs, seasons, days, years) are pertinent only to humans. The one that may seem not to belong is "seasons"—but here we must not think of seasons like summer and winter. The Hebrew word when it is used elsewhere designates the festival celebrations that are associated with the sowing season, the harvesting season and so on.[1]

Days four to six continue to be driven by the spoken word. This spoken word can easily be understood in connection to the establishment of functions. In the ancient Near East the cosmos is organized by the decrees of deity (reflected in the importance of the Tablet of Destiny). Genesis 1 also emphasizes the spoken decrees of the Creator, and these decrees initiate the functions and give the functionaries their roles. Such spoken decrees are also acts of creation. In ancient Mesopotamia the establishment of control attributes (Sumerian *me*) by decree and the functional aspects of the celestial bodies are combined in texts such as the Great Astrological Treatise:

> When An, Enlil, and Enki, the great gods,
> In their infallible counsel,
> Among the great laws *[me]* of heaven and earth.
> Had established the crescent of the moon,
> Which brought forth day, established the months
> and furnished the omens

> drawn from heaven and earth,
> This crescent shone in heaven,
> And one saw the stars shining in the highest heaven![2]

Similar interests and perspectives are attested throughout the ancient Near East.

Moving through day four, we should pause here a moment to comment on another verb associated with creative activity, *ʿāśâ*. This verb had been used in verse 7 ("God *made* the expanse"), and it is used again in day four, verse 16 ("God *made* two great lights"). It will be used again in day six for both animals (v. 25) and people (v. 26). It also shows up in some of the summary statements (Gen 2:2-4, variably as "made" or "done") and in Exodus 20:11 as a summary statement of the work of the seven days. While some may insist that this verb, at least, expresses a material perspective, we must be careful before jumping to such a conclusion. Any Hebrew lexicon will indicate that this verb covers the whole range, not only of "making" but also of "doing." Even in the summary statements in Genesis 2:2-4 the verb covers all the activities of the seven days, many of which clearly involve only doing, not making. It is true that this verb can be used for a material process, but it does not inherently refer to a material process. In Exodus 20, the discussion of the sabbath uses the same verb across verses 9-11. The phrases show a pattern: "In six days you shall *do* all your work . . . on the seventh . . . you shall not *do* any work . . . for in six days the Lord *did* the heavens and the earth [his work]." What does *doing* his work entail? If creation is his work, and creation is function oriented, then doing his work was accomplished by establishing functions.[3] This coincides with Genesis 2:2, which reports that God finished all the work he had been *doing* and rested from all the work of creating that he had *done*—all using the same verb.

On day four, God began with a decree (v. 14) that identified the functions of these celestial functionaries. Unlike the situation in the rest of the ancient Near East, these functionaries are nonpersonal entities. The text at least tacitly makes this point by referring to them as "lights" rather than by their names which coincided with the names of deities in the rest of the ancient Near East. Then he did the work so that they would govern as intended (v. 16). And finally he appointed them to their stations (v. 17). The conclusion is the familiar, "It was good" which, as we discussed last chapter, indicates that they are all prepared to function for the human beings that are soon going to be installed in their place.

DAY FIVE

In contrast to day four, where the functionaries were helping to accomplish the functions associated with the sphere which they inhabited, in day five the functionaries simply carry out their own functions in the cosmic space that they inhabit. The text addresses what they do (teem, fly) rather than the role they serve. But in the blessing God also gives them a function: to be fruitful and multiply. God created them capable of doing so, and it is their function to fill their respective realms.

Of particular interest is the specific attention paid to the "great creatures of the sea" in verse 21. Here the author returns to the verb he has not used since verse 1, *bārāʾ*, and which will only be used again in this chapter in verse 27. This use raises the significance of these creatures. In the ancient world the cosmic seas were populated with creatures that operated against the ordered system. Whether antithesis or enemy, they were viewed as threats to order, as they inhabited the region that was itself outside of the ordered system. This is the very reason why the author of Genesis would single them out for comment. Since there is no cosmic warfare or conquest in Genesis as is sometimes part of the ancient Near Eastern picture, the text indicates that these creatures are simply part of the ordered system, not enemies that had to be defeated and kept in check. In Genesis these creatures are fully under God's control.

Day Six

As with the creatures inhabiting cosmic space in day five, the animals inhabiting terrestrial space in day six are not functionaries that carry out the functions indicated in day three. Instead they carry out their own functions in that space. The text indicates their functions relative to their kind rather than functions relative to other inhabitants. They are viewed in their categories, and they reproduce after their own kind as part of the blessing of God. Their function is to reproduce and to fill the earth—this is what God made them to do. It is the wonder of creation that new generations of the same kinds of creatures are born from parent creatures. This is the same sort of marvel as the system that allows the plants to grow from seed.

One of the more intriguing elements in these verses is the subject and verb in verse 24 ("Let the land produce living creatures"). This is clearly not a scientific mode of expression, and the interpreter should not attempt to read in it scientific concepts. What would it refer to in an ancient Near Eastern context? As already mentioned, ancient Near Eastern texts do not often speak of the creation of animals, and when they do, it is generally a brief comment in passing. The closest statement to this one in Genesis comes from a work entitled The Exploits of Ninurta:

> Let its meadows produce herbs for you. Let its slopes produce honey and wine for you. Let its hillsides grow cedars, cypress, juniper and box for you. Let it make abundant for you ripe fruits, as a garden. Let the mountain supply you richly with divine perfumes. . . . Let the mountains make

> wild animals teem for you. Let the mountain increase the fecundity of quadrupeds for you.[4]

The role of the land or the mountains in producing animals does not give us material information as if this were some sort of spontaneous regeneration or a subtle indication of an evolutionary process. Rather the land and mountain are locations of origin. This is where animal life *comes from,* not what it is *produced from.* It is similar to a child today asking where babies come from. Rather than needing a description of sperm and egg in fertilization and conception, the child only needs to be told that babies come from hospitals or from their mothers.

HUMANITY

The difference when we get to the creation of people is that even as they function to populate the world (like fish, birds and animals), they also have a function relative to the rest of God's creatures, to subdue and rule. Not only that, but they have a function relative to God as they are in his image. They also have a function relative to each other as they are designated male and female. All of these show the functional orientation with no reference to the material at all. It could be claimed that the material aspect is picked up in Genesis 2, and we will discuss that in a separate section at the end of this chapter.

Among all of the functional elements referred to in Genesis 1:26-30, the image of God is the most important and is the focus of the section. All of the rest of creation functions in relationship to humankind, and humankind serves the rest of creation as God's vice regent. Among the many things that the image of God may signify and imply, one of them, and probably the main one, is that people are delegated a godlike role (function) in the world where he places them.

It has already been mentioned that whereas in the rest of the ancient world creation was set up to serve the gods, a theocentric view, in Genesis, creation is not set up for the benefit of God but for the benefit of humanity—an anthropocentric view. Thus we can say that humanity is the climax of the creation account. Another contrast between Genesis and the rest of the ancient Near East is that in the ancient Near East people are created to serve the gods by supplying their needs. That is, the role of people is to bring all of creation to deity—the focus is from inside creation out to the gods. In Genesis people represent God to the rest of creation. So the focus moves from the divine realm, through people, to the world around them. It would be like the difference between the employees in the plant who serve the company in the manufacturing process (like people in the ancient Near East) and the employees engaged in sales and marketing who represent the company to the outside world (like people in Genesis).

Materials for Humanity

Even though Genesis 1 mentions none of the materials or material processes for human origins, Genesis 2 appears to offer just such a description. Therefore we will step briefly out of our focus on Genesis 1 to address this issue.

Ancient Near Eastern texts contain numerous references to humans being created out of a variety of materials, and we find a great deal of continuity between those reports and the biblical text. This again tells us that Genesis is working within the normal conceptual framework of the ancient Near East rather than forging new scientific trails.

The materials or ingredients that are attested in the ancient Near East are tears of a god (Egypt), blood of a god (Atrahasis), and the most common, clay (both Egypt and Mesopotamia). These ingredients are offered as common to all of humanity since the ancient Near Eastern texts only deal with the mass of humanity being created rather than an individual or a couple as in Genesis. This is an important difference as Adam and Eve are treated as individuals in chapters 4 and 5. This individual identity, however, does not change the significance of the reference to the materials in Genesis 2. The fact that the ancient Near East uses the same sorts of materials to describe all of humanity indicates that the materials have archetypal significance. Unlike a prototype (which is an original item that serves as a model for later production), an archetype serves as a representative for all others in the class and defines the class. So when the ancient Near Eastern texts speak of people being created from clay or the blood of a slain deity, they are not talking about just one individual, but are addressing the nature of all humanity.

This archetypal understanding applies also to Genesis 2. An individual named Adam is not the only human being made of the dust of the earth, for as Genesis 3:19 indicates, "Dust you are and to dust you will return." This is true of all humans, men and women. It is an archetypal feature that describes us all. It is not a statement of chemical composition nor is it describing a material process by which each and every human being is made. The dust is an archetypal feature and therefore cannot be viewed as a material ingredient. It is indicative of human destiny and mortality, and therefore is a functional comment, not a material one.

The situation is no different with the creation of woman. Being drawn from the side of man has an archetypal significance, not an anatomical one. This is the very aspect that the text draws out when it identifies the significance of the detail: "For this reason a man will leave his father and mother and be united to his wife, and they will become one flesh" (Gen 2:24). This is true of all mankind and all womankind. Womankind is archetypally made from the side of mankind. Again we can see that this is a functional discussion, not a material one. After chapter five of Genesis, Adam and Eve are never again mentioned in the Old Testament except in the opening genealogy in Chronicles. In the New Testament, the authors regularly treat Adam and Eve in archetypal terms.[5]

Given these observations, we might conclude that Genesis does not have the same level of interest in the material origins of the first humans as we do. It focuses its attention on the archetypal origins of humanity, mankind and womankind. This interest is part of functional origins. Humankind is connected to the ground from which we are drawn. Womankind is connected to mankind from whom she is drawn. In both male and female forms, humankind is connected to God in whose image all are made. As such they have the privilege of procreation, the role of subduing and ruling, and a status in the garden serving sacred space (Gen 2:15). All of these, even the last, were designed to be true of all human beings. Neither the materials nor the roles are descriptive only of the first individuals. This creation account gives people their identity and specifies their connectivity to everything around them.

Summary

In days four to six the functionaries of the cosmos are installed in their appropriate positions and given their appropriate roles. Using the company analogy, they are assigned their offices (cubicles), told to whom they will report, and thus given an idea of their place in the company. Their workday is determined by the clock, and they are expected to be productive. Foremen have been put in place, and the plant is now ready for operation. But before the company is ready to operate, the owner is going to arrive and move into his office.

PROPOSITION 7

Divine Rest Is in a Temple

In the traditional view that Genesis 1 is an account of material origins, day seven is mystifying. It appears to be nothing more than an afterthought with theological concerns about Israelites observing the sabbath—an appendix, a postscript, a tack on.

In contrast, a reader from the ancient world would know immediately what was going on and recognize the role of day seven. Without hesitation the ancient reader would conclude that this is a temple text and that day seven is the most important of the seven days. In a material account day seven would have little role, but in a functional account, as we will see, it is the true climax without which nothing else would make any sense or have any meaning.

How could reactions be so different? The difference is the piece of information that everyone knew in the ancient world and to which most modern readers are totally oblivious: Deity rests in a temple, and only in a temple. This is what temples were built for. We might even say that this is what a temple is—a place for divine rest. Perhaps even more significant, in some texts the construction of a temple is associated with cosmic creation.

What does divine rest entail? Most of us think of rest as disengagement from the cares, worries and tasks of life. What comes to mind is sleeping in or taking an afternoon nap. But in the ancient world rest is what results when a crisis has been resolved or when stability has been achieved, when things have "settled down." Consequently normal routines can be established and enjoyed. For deity this means that the normal operations of the cosmos can be undertaken. This is more a matter of engagement without obstacles rather than disengagement without responsibilities.

Before we proceed, it is important to look at the terminology used by the author. The Hebrew verb *šābat* (Gen 2:2) from which our term "sabbath" is derived has the basic meaning of "ceasing" (cf. Josh 5:12; Job 32:1). Semantically it refers to the completion of certain activity with which one had been occupied. This cessation leads into a new state which is described by another set of words, the verb *nûḥa* and its associated noun, *mĕnûḥâ*. The verb involves entering a position of safety, security or stability and the noun refers to the place where that is found. The verb *šābat* describes a transition into the activity or inactivity of *nûḥa*. We know that when God rests (ceases, *šābat*) on the seventh day in Genesis 2, he also transitions into the condition of stability *(nûḥa)* because that is the terminology used in Exodus 20:11. The only other occurrence of the verb *šābat* with God as the subject is in Exodus 31:17.[1] The most important verses to draw all of this information together are found in Psalm 132:7-8, 13-14.

> Let us go to his dwelling place;
> let us worship at his footstool—
> "arise, O Lord, and come to your resting place,
> you and the ark of your might."
>
> For the Lord has chosen Zion,
> he has desired it for his dwelling:
> "This is my resting place for ever and ever;

here I will sit enthroned, for I have desired it."

Here the "dwelling place" of God translates a term that describes the tabernacle and temple, and this is where his footstool (the ark) is located. This also shows that the text is referring to his dwelling place as his throne room and the place of his rule (because of the footstool). In verse 8 the "footstool" is paralleled by the ark, and the temple ("dwelling place") is paralleled with "resting place" *(mĕnûḥâ)*. This demonstrates that the temple is the place where he rests. In verse 13 the text again refers to his dwelling in Zion, thus referring to the temple. Then verse 14 uses "resting place" *(mĕnûḥâ)* again identifying it as the place where he is enthroned. Thus, this Psalm pulls together the ideas of divine rest, temple and enthronement. God's "ceasing" *(šābat)* on the seventh day in Genesis 2:2 leads to his "rest" *(nûḥa)*, associated with the seventh day in Exodus 20:11. His "rest" is located in his "resting place" *(mĕnûḥâ)* in Psalm 132, which also identifies it as the temple from which he rules. After creation, God takes up his rest and rules from his residence. This is not new theology for the ancient world—it is what all peoples understood about their gods and their temples.

In the Old Testament the idea that rest involves engagement in the normal activities that can be carried out when stability has been achieved can be seen in the passages where God talks of giving Israel rest in the land:

> But you will cross the Jordan and settle in the land the Lord your God is giving you as an inheritance and he will give you rest from all your enemies around you so that you will live in safety. (Deut 12:10; cf. Josh 21:44; 23:1)

Although security and stability might allow one to relax, more importantly it allows life to resume its normal routines. When Israel's enemies no longer threaten, they can go about their lives: planting and harvesting, buying and selling, raising their families and serving their God.

In the same way, a temple is built in the ancient world so that deity can have a center for his rule. The temple is the residence and palace of the gods. Like the American White House, it is the hub of authority and control. It is where the work of running the country takes place. When a newly elected president looks forward to taking up his residence in the White House, it is not simply so he can kick off his shoes and snooze in the Lincoln bedroom. It is so he can begin the work of running the country. Thus in ancient terms the president "takes up his rest" in the White House. This is far from relaxation. The turmoil and uncertainty of the election is over, and now he can settle down to the important business at hand.

The role of the temple in the ancient world is not primarily a place for people to gather in worship like modern churches. It is a place for the deity—sacred space. It is his home, but more importantly his headquarters—the control room. When the deity rests in the temple it means that he is taking command, that he is mounting to his throne to assume his rightful place and his proper role.

In ancient Near Eastern literature this concept appears early and often. One of the earliest available Sumerian literary pieces is the Temple Hymn of Keš:

> House …… inspiring great awe, called with a mighty name by An; house …… whose fate is grandly determined by the Great Mountain Enlil! House of the Anuna gods possessing great power, which gives wisdom to the people; house, reposeful dwelling of the great gods! House, which was planned together with the plans of heaven and earth, ……

> with the pure divine powers; house which underpins the Land and supports the shrines![2]

In this hymn we can see the idea that the temple is a place of rest ("reposeful dwelling"), that it is central in functional creation ("planned together with the plans of heaven and earth"), and that it is the place from which control is exercised ("underpins the land").

In the famous Babylonian creation epic, *Enuma Elish,* the work of creation by Marduk is followed by the building of a temple for him. Note the following: The gods give Marduk kingship (5.113), and Marduk responds with the statement, "Below the firmament, whose grounding I have made firm, A house I shall build, let it be the abode of my pleasure. Within it I shall establish its holy place, I shall appoint my holy chambers, I shall establish my kingship" (5.121-24). This place is to be the "stopping place" of the gods (5.138). After humankind is created at the beginning of tablet six and the gods are given their responsibilities, the head gods make a declaration: "We will make a shrine, whose name will be a byword, your chamber that shall be our stopping place, we shall find rest therein" (6.51-52).[3] These sections demonstrate the close relationship between creation (cosmic and human), organization of the cosmos, rest, temple and rule.

God's resting in Genesis 1 does not specifically describe his engagement of the controls, but it describes the opportunity to do so. He can disengage from the set-up tasks and begin regular operations. It would be similar to getting a new computer and spending focused time setting it up (placing the equipment, connecting the wires, installing the software). After all of those tasks were done, you would disengage from that process, mostly so you could now engage in the new tasks of actually using the computer. That is what it had been set up for.[4]

Sometimes people have raised the question, What did God do on the eighth day? In the view being presented here, on the eighth day, and on every day since then, he is in the control room from where he runs the cosmos that he set up. This is the ongoing work of creation. When we thought of Genesis 1 as an account of material origins, creation became an action in the past that is over and done with. God made objects and now the cosmos exists (materially). Viewing Genesis 1 as an account of functional origins offers more opportunity for understanding that God's creative work continues (more about this in chapter 15).

Now that we have been given the interpretive key from the world of the ancient Near East (and verified in other portions of the Bible as well) that divine rest is in a temple, we can begin to unpack the significance of this information for further understanding Genesis 1. What are the implications of identifying Genesis 1 as a "temple text"? What temple is being referred to, and what does that tell us about Genesis 1 and about theology? These are the topics to be addressed in the next several chapters.

Technical Support

Andreasen, N.-E. *The Old Testament Sabbath: A Tradition-Historical Investigation.* SBL Dissertation Series. Missoula, Mont.: Society of Biblical Literature, 1972.

Laansma, J. *I Will Give You Rest.* Tübingen: Mohr, 1997.

PROPOSITION 8

The Cosmos Is a Temple

IN SOME OF THE ANCIENT Near Eastern texts, a temple is built as a conclusion to cosmic creation. But typically these are distinct, though related acts. The natural association between them is that the creative acts are expressions of authority, and the temple is the place where authority will continue to be exercised. Beyond this textual and ideological association, we can see that texts link creation and temple building by noting the absence of temples along with the absence of cosmic order as they recount the acts of creation. Thus the absence of a temple was sometimes part of the description of the precosmic condition. This is clearest in the preamble to a prayer that concerns the founding of Eridu:[1]

> No holy house, no house of the gods, had been built in a pure place;
> No reed had come forth, no tree had been created;
> No brick had been laid, no brickmold had been created;
> No house had been built, no city had been created;
> No city had been built, no settlement had been founded;
> Nippur had not been built, Ekur had not been created;
> Uruk had not been built, Eanna had not been created;

> The depths had not been built, Eridu had not been created;
> No holy house, no house of the gods, no dwelling for them had been created.
> All the world was sea,
> The spring in the midst of the sea was only a channel,
> Then was Eridu built, Esagila was created.[2]

Then Marduk settles the gods into their dwelling places, creates people and animals, and sets up the Tigris and Euphrates.

In a prayer to dedicate the foundation brick of a temple it is obvious that the cosmos and temple were conceived together and thus are virtually simultaneous in their origins.

> When Anu, Enlil, and Ea had a (first) idea of heaven and earth,
> They found a wise means of providing support of the gods:
> They prepared, in the land, a pleasant dwelling,
> And the gods were installed (?) in this dwelling:
> Their principal temple.[3]

This close connection between cosmic origins and temple building reinforces the idea across the ancient Near East that the temples were considered primordial and that cosmic origins at times were defined in terms of a temple element. It is important to reiterate that I am not suggesting that the Israelites are borrowing from these ancient literatures. Instead the literatures show how people thought in the ancient world, and as we examine Genesis, we can see that Israelites thought in similar ways.

We can draw the connection between temple and cosmos more tightly when we observe that temples in the ancient world were considered symbols of the cosmos. The biblical text as well as the literature of the ancient Near East makes this clear. Ancient Near Eastern evidence comes from a variety of cultures and sources.

First, temples had cosmic descriptions in the ancient world. The earliest example is in the Sumerian Temple Hymn of Kes, one of the oldest pieces of literature known.[4]

> House Keš, platform of the Land, important fierce bull!
> Growing as high as the hills, embracing the heavens,
> Growing as high as E-kur, lifting its head among the mountains!
> Rooted in the Abzu, verdant like the mountains!

The Sumerian text of Gudea's construction of a temple shows the temple serving a cosmic function. Toward the end of Cylinder B, the god Ningirsu, speaking to Gudea, suggests that it is the temple that separates heaven and earth, thus associating it with that most primordial act of creation:

> [Gu]dea, you were building my [house] for me,
> And were having [the offices] performed to perfection [for me],
> You had [my house] shine for me
> Like Utu in [heaven's midst],
> Separating. Like a lofty foothill range,
> Heaven from earth.[5]

Many of the names given to temples in the ancient world also indicate their cosmic role. Among the dozens of possible examples, note especially the temple Esharra ("House of the Cosmos") and Etemenanki ("House of the Foundation Platform Between Heaven and Earth").

In Egypt temples were regarded as having been built where the primeval hillock of land first emerged from the cosmic waters.

> The temple recalled a mythical place, the primeval mound. It stood on the first soil that emerged from the primeval waters, on which the creator god stood to begin his work of creation. Through a long chain of ongoing renewals, the present temple was the direct descendant of the original sanctuary that

> the creator god himself had erected on the primeval mound. An origin myth connecting the structure with creation is associated with each of the larger late temples.[6]

Both Sumerian and Egyptian texts identify the temple as the place from which the sun rises: "Your interior is where the sun rises, endowed with wide-spreading plenty."[7] The Egyptian temples served as models of the cosmos in which the floor represented the earth and the ceiling represented the sky. Columns and wall decorations represented plant life. Jan Assmann, presenting this imagery, concludes that the temple "*was* the world that the omnipresent god filled to its limits."[8] Indeed, the temple is, for all intents and purposes, the cosmos.[9] This interrelationship makes it possible for the temple to be the center from which order in the cosmos is maintained.[10]

In the biblical text the descriptions of the tabernacle and temple contain many transparent connections to the cosmos. This connection was explicitly recognized as early as the second century A.D. in the writings of the Jewish historian Josephus, who says of the tabernacle: "every one of these objects is intended to recall and represent the universe."[11] In the outer courtyard were representations of various aspects of cosmic geography. Most important are the water basin, which 1 Kings 7:23-26 designates "sea," and the bronze pillars, described in 1 Kings 7:15-22, which perhaps represented the pillars of the earth. The horizontal axis in the temple was arranged in the same order as the vertical axis in the cosmos. From the courtyard, which contained the elements outside the organized cosmos (cosmic waters and pillars of the earth), one would move into the organized cosmos as he entered the antechamber. Here were the Menorah, the Table of Bread and the incense altar. In the Pentateuch's descriptions of the tabernacle, the lamp and its olive oil are provided for "light" (especially Ex 25:6; 35:14; Num 4:9). This word for light is the same word used to describe the celestial bodies in day four (rather than calling them sun and moon). As the Menorah represented the light provided by God, the "Bread of the Presence" (Ex 25:30) represented food provided by God. The altar of incense provided a sweet-smelling cloud across the face of the veil that separated the two chambers. If we transpose from the horizontal axis to the vertical, the veil separated the earthly sphere, with its functions, from the heavenly sphere, where God dwells. This latter was represented in the holy of holies, where the footstool of the throne of God (the ark) was placed. Thus the veil served the same symbolic function as the firmament. To review then, the courtyard represented the cosmic spheres outside of the organized cosmos (sea and pillars). The antechamber held the representations of lights and food. The veil separated the heavens and earth—the place of God's presence from the place of human habitation.[12]

Scholars have also recognized that the temple and tabernacle contain a lot of imagery from the Garden of Eden. They note that gardens commonly adjoined sacred space in the ancient world. Furthermore the imagery of fertile waters flowing from the presence of the deity to bring abundance to the earth is a well-known image.

> The garden of Eden is not viewed by the author of Genesis simply as a piece of Mesopotamian farmland, but as an archetypal sanctuary, that is a place where God dwells and where man should worship him. Many of the features of the garden may also be found in later sanctuaries particularly the tabernacle or Jerusalem temple. These parallels suggest that the garden itself is understood as a sort of sanctuary.[13]

So the waters flowing through the garden in Genesis 2 are

paralleled by the waters flowing from the temple in Ezekiel 47:1-12 (cf. Ps 46:4; Zech 14:8; Rev 22:1-2). This is one of the most common images in the iconography of the ancient world.[14] Consequently we may conclude that the Garden of Eden was sacred space and the temple/tabernacle contained imagery of the garden and the cosmos. All the ideas are interlinked. The temple is a microcosm, and Eden is represented in the antechamber that serves as sacred space adjoining the Presence of God as an archetypal sanctuary.

From the idea that the temple was considered a mini cosmos, it is easy to move to the idea that the cosmos could be viewed as a temple. This is more difficult to document in the ancient world because of the polytheistic nature of their religion. If the whole cosmos were viewed as a single temple, which god would it belong to? Where would temples of the other gods be? Nevertheless it can still be affirmed that creation texts can and do follow the model of temple-building texts, in this way at least likening the cosmos to a temple.[15]

In the Old Testament, polytheism would not interfere with the association of cosmos and temple, and indeed the connection is made. Isaiah 66:1-2 is the clearest text.

> This is what the Lord says:
> "Heaven is my throne,
> and the earth is my footstool.
> Where is the house you will build for me?
> Where will my resting place be?
> Has not my hand made all these things,
> and so they came into being?"
> declares the Lord.

Here we can see the elements of a cosmos-sized temple, a connection between temple and rest, and a connection between creation and temple. This in itself is sufficient to see that the cosmos can be viewed as a temple. That is precisely what we are proposing as the premise of Genesis 1: that it should be understood as an account of functional origins of the cosmos as a temple. Other passages in the Old Testament that suggest the cosmos be viewed as a temple include 1 Kings 8:27, where in his prayer dedicating the temple, Solomon says, "But will God really dwell on earth? The heavens, even the highest heaven, cannot contain you. How much less this temple that I have built?" In another, Isaiah 6:3, the seraphim chant, "Holy, holy, holy, is the Lord Almighty, the whole earth is full of his glory." The "glory" that the earth is full of is the same as that which comes and takes up residence in the holy of holies in Exodus 40:34.[16]

This chapter has given evidence for the following:

1. In the Bible and in the ancient Near East the temple is viewed as a microcosm.
2. The temple is designed with the imagery of the cosmos.
3. The temple is related to the functions of the cosmos.
4. The creation of the temple is parallel to the creation of the cosmos.
5. In the Bible the cosmos can be viewed as a temple.

When this information is combined with the discoveries of the last chapter—that deity rests in a temple, and that therefore Genesis 1 would be viewed as a temple text—we gain a different perspective on the nature of the Genesis creation account. Genesis 1 can now be seen as a creation account focusing on the cosmos as a temple. It is describing the creation of the cosmic temple with all of its functions and with God dwelling in its midst. This is what makes day seven so significant, because without God taking up his dwelling in its midst, the (cosmic) temple does not exist. The most central truth to the creation account is that this world is a place for God's presence. Though all of the functions are anthropocentric, meeting the needs of humanity, the cosmic temple is theocentric, with God's presence serving as the defining element of existence. This represents a change that has taken place over the seven days. Prior to day one, God's spirit was active over the nonfunctional cosmos; God was involved but had not yet taken up his residence. The establishment of the functional cosmic temple is effectuated by God taking up his residence on day seven. This gives us a before/after view of God's role.

Technical Support

Beale, G. K. *The Temple and the Church's Mission: A Biblical Theology of the Dwelling Place of God.* Downers Grove, Ill.: InterVarsity Press, 2004.

Hurowitz, Victor. *I Have Built You an Exalted House: Temple Building in the Bible in Light of Mesopotamian and Northwest Semitic Writings.* Journal for the Study of the Old Testament Supplement Series 115. Sheffield, U.K.: JSOT Press, 1992.

———. "Yhwh's Exalted House—Aspects of the Design and Symbolism of Solomon's Temple," in *Temple and Worship in Biblical Israel,* pp. 63-110. Edited by J. Day. New York: T & T Clark, 2005.

Levenson, Jon. "The Temple and the World," *Journal of Religion* 64 (1984): 275-98.

———. *Creation and the Persistence of Evil.* Princeton, N.J.: Princeton University Press, 1988.

Lundquist, J. "What Is a Temple? A Preliminary Typology." In *The Quest for the Kingdom of God,* pp. 205-19. Winona Lake, Ind.: Eisenbrauns, 1983.

Weinfeld, Moshe. "Sabbath, Temple, and the Enthronement of the Lord—The Problem of the Sitz im Leben of Genesis 1.1—

2.3." In *Mélanges bibliques et orientaux en l'honneur de M. Henri Cazelles,* edited by A. Caquot and M. Delcor, pp. 501-12. Alter Orient und Altes Testament 212. Neukirchen-Vluyn: Neukirchener; Kevelaer: Butzon & Bercker, 1981.

Wenham, Gordon J. "Sanctuary Symbolism in the Garden of Eden Story." In *I Studied Inscriptions from Before the Flood,* edited by R. S. Hess and D. Toshio Tsumura, pp. 399-404. Sources for Biblical and Theological Study 4. Winona Lake, Ind.: Eisenbrauns, 1994. Reprinted from *Proceedings of the Ninth World Congress of Jewish Studies, Division A: The Period of the Bible,* pp. 19-25. Jerusalem: World Union of Jewish Studies, 1986.

PROPOSITION 9

The Seven Days of Genesis 1 Relate to the Cosmic Temple Inauguration

THE RELATIONSHIP BETWEEN cosmos and temple in the Bible and in the ancient world, and particularly the common connection between the two in creation texts suggests that we should think of Genesis 1 in relation to a cosmic temple. This is further confirmed by the divine rest on the seventh day, since divine rest takes place in temples. These ideas should lead us to investigate what other elements of Genesis 1 might be affected by thinking in temple terms.

First in line is the curious fact that the number seven appears so pervasively in temple accounts in the ancient world and in the Bible.[1] Thus the seven days of the Genesis account of origins has a familiarity that can hardly be coincidental and tells us something about the seven-day structure in Genesis 1 that we did not know before and that is not transparent to modern readers. That is, if Genesis 1 is a temple text, the seven days may be understood in relation to some aspect of temple inauguration. What would days of inauguration have to do with creation? What is the connection? If Genesis 1 were an account of material origins, there would be no connection at all. But as an account of functional origins, creation and temple inauguration fit hand in glove. Given the relationship of the temple and the cosmos, the creation of one is also the creation of the other. The temple is made functional in the inauguration ceremonies, and therefore the temple is created in the inauguration ceremony. So also the cosmic temple would be made functional (created) in an inauguration ceremony.

We must draw an important distinction between the *building* of a temple and the *creation* of a temple. When we look again at the account of Solomon's temple we see that he took seven years to build it (1 Kings 6:37-38). Most of this time was spent on what may be called the "material phase." The stone was quarried and shaped, the precious metals were mined, the furniture built, the cedar acquired and shipped and shaped, the veils sewn, the doors carved, the priestly vestments made and so on. When all of this was done, did the temple exist? Certainly not. Because a temple is not simply an aggregate of fine materials subjected to expert craftsmanship. The temple uses that which is material, but the temple is not material. If God is not in it, it is not a temple. If rituals are not being performed by a serving priesthood, it is not a temple. If those elements are not in place, the temple does not exist in any meaningful way. A person does not exist if only represented by their corpse. It is the inauguration ceremony that transforms a pile of lumber, stone, gold and cloth into a temple.

What happens in a temple inauguration to cause this transformation? We have many inauguration texts from the ancient world, the most detailed being the dedication of the temple of Ningirsu by Gudea about 2100 B.C. One of the first things to note is that at the inauguration the "destiny" and the powers of the temple are assigned (Gudea B.i.3; xiii.6). This is the ultimate function-giving act in the ancient world. Likewise the roles of the functionaries are proclaimed and they are installed.[2]

> To guide aright the hand of the one who does righteousness;
> To put the wood (neck stock) on the neck of the one who does evil;
> To keep the temple true; to keep the temple good;
> To give instructions to his city, the sanctuary Girsu;
> To set up the throne of decreeing destiny;
> To put into the hand the scepter of prolonged days.[3]

In short, by naming the functions and installing the functionaries, and finally by deity entering his resting place, the temple comes into existence—it is created in the inauguration ceremony.

A good biblical example can be seen in the tabernacle account in Exodus 35-39, which concerns the material phase. Exodus 39:32 gives the report on the material phase: "So all the work on the tabernacle, the Tent of Meeting, was completed. The Israelites did everything just as the Lord commanded Moses." In Exodus 39:43, after they have brought everything to Moses, he inspects it, and judges it worthy of blessing. Exodus 40 describes the inauguration—this is the creation of the tabernacle. The chapter reports everything being put in its place, anointed and consecrated (Ex 40:9-16). When all of this is done, the inauguration is completed by the glory of the Lord filling the tabernacle (Ex 40:34). In Exodus we are not told whether all of this was done in one day or over several days, but we do see that it is done in connection with the New Year (Ex 40:2, 17).

Inauguration ceremonies are described in the Old Testament with various levels of detail, including the activities of cultic ritual for consecration and sacrifices that initiate the operation of the sacred space. The Hebrew term is *ḥănukkâ* (see Num 7:10-11, 84, 88; 1 Kings 8:63; 2 Chron 7:5; note also Ps 30). The dedication is the celebration of the people that typically follows, though perhaps at times overlaps with, the inauguration. In the account of the construction of Solomon's Temple the inauguration includes a seven-day dedication to which is added a seven-day feast/banquet (1 Kings 8:65; 2 Chron 7:9). Solomon's dedicatory prayer proclaims the functions of the temple:

- place for seeking forgiveness (1 Kings 8:30)
- place for oath swearing (1 Kings 8:31-32)
- place for supplication when defeated (1 Kings 8:33-34)
- place for supplication when faced with drought/famine/blight (1 Kings 8:35-40)
- place for the alien to pray (1 Kings 8:41-43)
- place for petition for victory (1 Kings 8:44-45)

In the ancient world the building or restoration of a temple was one of the most notable accomplishments that a ruler could undertake. It was believed to bring the favor of the god, to bring benefits to the city and to bring order to the cosmos. Of course when the temple project was complete there were inauguration activities, consecration, cultic acts, dedication and great public ceremonies. But that was not the end of it. Temple inauguration could also be reenacted on a yearly basis, and pieces of literature like the Sumerian Temple Hymns may have served as the liturgy for such annual celebrations. In Babylon one of the most well-known festivals was the Akitu festival, often celebrated in connection with the New Year, which reinstalled the deity in the temple and reasserted the king's selection by the gods. The Babylonian creation epic, *Enuma Elish,* was read in connection with this festival as it recounts the god Marduk's ascension to the head of the gods and his building of the temple along with his acts of creation.

Long controversy has existed as to whether Israel practiced similar enthronement festivals or New Year celebrations that reaffirmed creation, temple presence and royal election. The Bible contains no clear evidence of such festivals, but some see hints that they think point that direction. It would be no surprise if they had such a festival and would be theologically and culturally appropriate. Moshe Weinfeld has suggested that Genesis 1 could have served very effectively as the liturgy of such a festival,[4] and the suggestion has much to commend it both textually and culturally, though definitive evidence is lacking. In this way of thinking, Genesis 1 would be a recounting of the functional origins of the cosmos as a temple that was rehearsed yearly to celebrate God's creation and enthronement in the temple.

In this view of Genesis 1, it is evident that the nature of the days takes on a much less significant role than has normally been the case in views that focus on material creation, in that they no longer have any connection to the material age of the earth. These are seven twenty-four-hour days. This has always been the best reading of the Hebrew text. Those who have tried to alleviate the tension for the age of the earth commonly suggested that the days should be understood as long eras (the day-age view). This has never been convincing. The evidence used by the proponents of the day-age view is that the word translated "day" *(yôm)* is often a longer period of time, and they chose that meaning for the word in Genesis 1. The first problem with this approach is that the examples generally used of *yôm* referring to an extended period of time are examples in which the word is being used idiomatically: "in that day." This is a problem because words often take on specialized meaning in idiomatic expressions. So in Hebrew, the phrase "in that day" is simply a way for Hebrew to say "when." The word *yôm* cannot be removed from that expression and still carry the meaning that it has in the expression. Second, if it could be established that the word *yôm* could refer to a longer period of time, the interpreter would still have the responsibility for determining which meaning the author intended in the passage. Word meanings cannot be chosen as if we were in a cafeteria taking whatever we like. Third, the attempt to read long periods of time is clearly a concordist resort,[5] which will be discussed in chapter eleven.

The day-age theory and others that attempt to mitigate the force of the seven days do so because they see no way to reconcile seven twenty-four-hour days of material creation with the evidence from science that the earth and the universe are very old. They seek a solution in trying to stretch the meaning of *yôm,* whereas we propose that once we understand the nature of the creation account, there is no longer any need to stretch *yôm.*

In summary, we have suggested that the seven days are not given as the period of time over which the material cosmos came into existence, but the period of time devoted to the inauguration of the functions of the cosmic temple, and perhaps also its annual reenactment. It is not the material phase of temple construction that represents the creation of the temple; it is the inauguration of the functions and the entrance of the presence of God to take up his rest that creates the temple. Genesis 1 focuses on the creation of the (cosmic) temple, not the material phase of preparation. In the next chapter we will track the implications of the idea that the seven days are not related to the material phase of creation.

PROPOSITION 10

The Seven Days of Genesis 1 Do Not Concern Material Origins

Previous chapters proposed that Genesis 1 is not an account of material origins but an account of functional origins, specifically focusing on the functioning of the cosmos as God's temple. In the last chapter we identified the seven days of creation as literal twenty-four-hour days associated with the inauguration of the cosmic temple—its actual creation, accomplished by proclaiming its functions, installing its functionaries, and, most importantly, becoming the place of God's residence.

One of the most common questions about this view comes from those who are struggling with the worldview shift from material orientation to functional orientation (a difficult jump for all of us). In a last effort to cling to a material perspective, they ask, why can't it be both? It is easy to see the functional orientation of the account, but does the material aspect have to be eliminated altogether?

In answer to this question, if we say that the text includes a material element alongside the functional, this view has to be demonstrated, not just retained because it is the perspective most familiar to us. The comfort of our traditional worldview is an insufficient basis for such a conclusion. We must be led by the text. A material interest cannot be assumed by default, it must be demonstrated, and we must ask ourselves why we are so interested in seeing the account in material terms. In previous chapters I have proposed the following:

- The nature of the governing verb (*bārāʾ*, "create") is functional.
- The context is functional (it starts with a nonfunctional world in Gen 1:2 and comes back to a functional description of creation after the flood in Gen 8:22).[1]
- The cultural context is functional (ancient Near Eastern literature).
- The theology is functional (cosmic temple).

These provide some significant evidences of the functional perspective.

If we turn our attention to the possible evidences for the material interests of the account we find significant obstacles:

- Of the seven days, three have no statement of creation of any material component (days 1, 3 and 7).
- Day two has a potentially material component (the firmament, *rāqîʿa*), but no one believes there is actually something material there—no solid construction holds back the upper waters. If the account is material as well as functional we then find ourselves with the problem of trying to explain the material creation of something that does not exist. The word *rāqîʿa* had a meaning to Israelites as referring to a very specific object in their cosmic geography. If this were a legitimate material account, then we would be obliged to find something solid up there (not just change the word to mean something else as concordists tend to do). In the functional approach, this compo-

nent of Old World science addresses the function of weather, described in terms that they would understand.

- Days four and six have material components, but the text explicitly deals with them only on the functional level (celestial bodies for signs, seasons, days and years; human beings in God's image, male and female, with the task to subdue and rule).
- This leaves only day five in discussion, where functions are mentioned (e.g., let them swarm) and the verb *bārāʾ* is again used.[2] As a result, it is difficult to sustain a case that the account is interested in material origins if one does not already come with that presupposition.

If the seven days refer to the seven days of cosmic temple inauguration, days that concern origins of functions not material, then the seven days and Genesis 1 as a whole have nothing to contribute to the discussion of the age of the earth. This is not a conclusion designed to accommodate science—it was drawn from an analysis and interpretation of the biblical text of Genesis in its ancient environment. The point is *not* that the biblical text therefore supports an old earth, but simply that there is no biblical position on the age of the earth. If it were to turn out that the earth is young, so be it. But most people who seek to defend a young-earth view do so because they believe that the Bible obligates them to such a defense. I admire the fact that believers are willing to take unpopular positions and investigate all sorts of alternatives in an attempt to defend the reputation of the biblical text. But if the biblical text does not demand a young earth there would be little impetus or evidence to offer such a suggestion.

If there is no biblical information concerning the age of the material cosmos, then, as people who take the Bible seriously, we have nothing to defend on that count and can consider the options that science has to offer. Some scientific theories may end up being correct and others may be replaced by new thinking. We need not defend the reigning paradigm in science about the age of the earth if we have scientific reservations, but we are under no compulsion to stand against a scientific view of an old earth because of what the Bible teaches.[3]

One of the sad statistics of the last 150 years is that increasing numbers of young people who were raised in the environment of a biblical faith began to pursue education and careers in the sciences and found themselves conflicted as they tried to sort out the claims of science and the claims of the faith they had been taught. It seems to many that they have to make a choice: either believe the Bible and hold to a young earth, or abandon the Bible because of the persuasiveness of the case for an old earth. The good news is that we do not have to make such a choice. The Bible does not call for a young earth. Biblical faith need not be abandoned if one concludes from the scientific evidence that the earth is old.

At this point a very clear statement must be made: *Viewing Genesis 1 as an account of functional origins of the cosmos as temple does* not *in any way suggest or imply that God was uninvolved in material origins—it only contends that Genesis 1 is not that story.* To the author and audience of Genesis, material origins were simply not a priority. To that audience, however, it would likewise have been unthinkable that God was somehow uninvolved in the material origins of creation. Hence there wouldn't have been any need to stress a material creation account with God depicted as centrally involved in material aspects of creation. We can understand this issue of focused interests through any number of analogies from our own world as we indicated in chapter two with the examples of a company and a computer. Many situations in our experience interest us on the functional level while they generate no curiosity at all about the material aspect.

Our affirmation of God's creation of the material cosmos is supported by theological logic as well as by occasional New Testament references. By New Testament times there was already a growing interest in material aspects and so also a greater likelihood that texts would address material questions. Speaking of Christ, Paul affirms, "For by him all things were created: things in heaven and on earth, visible and invisible, whether thrones or powers of rulers or authorities; all things were created by him and for him. He is before all things, and in him all things hold together" (Col 1:16-17). This statement can certainly be understood to include both the material and the functional. Hebrews 1:2 is less explicit as it affirms that the Son is appointed the heir of all things and that through him God made the "universe." Here it must be noted that the word translated "universe" is *aiōnas*, not *kosmos*—thus more aptly referring to the ages of history than to the material world (the same in Heb 11:3).

The theological point is that whatever exists, be it material or functional, God made it. But from there our task as interpreters is to evaluate individual texts to see what aspect of God's creation they discuss.

Finally we need to address the question of what actually happens in the seven days. What would a comparison of the "before" and "after" pictures look like? What would an observer see if able to observe the process of these seven days? On these we can only speculate, but I will try to explore the implications of this view.

The functional view understands the functions to be decreed by God to serve the purposes of humanity, who has been made in his image. The main elements lacking in the "before" picture are therefore humanity in God's image and God's presence in his cosmic temple. Without those two ingredients the cosmos would be considered nonfunctional and therefore nonexistent. The material phase nonetheless could have been under development for long eras and could in that case correspond with the descriptions of the prehistoric ages as science has uncovered them for us. There would be no reason to think that the sun had not been shining, plants had not been growing, or animals had not been present.[4] These were like the rehearsals leading up to a performance of a play. The rehearsals are preparatory and necessary, but they are not the play. They find their meaning only when the audience is present. It is then that the play exists, and it is for them that the play exists.

In the "after" picture the cosmos is now not only the handiwork of God (since he was responsible for the material phase all along, whenever it took place), but it also becomes God's residence—the place he has chosen and prepared for his presence to rest. People have been granted the image of God and now serve him as vice regents in the world that has been made for them. Again it is instructive to invoke the analogy of the temple before and after its inauguration. After priests have been installed and God has entered, it is finally a fully functioning temple—it exists only by virtue of those aspects.

What would a college be without students? Without administration and faculty? Without courses? We could talk about the origins of the college when it first opened its doors, enrolled students for the first time, hired faculty, designed courses and offered them and so on. In another sense this process is reenacted year by year as students return (or are newly enrolled), faculty again inhabit their offices, courses are offered. Anyone in academics knows the difference between the empty feel of campus during the summer compared to the energy of a new semester beginning.

Before the college existed, there would have been a material "construction" phase. What a mess! Partially built buildings, construction equipment, torn up ground and so forth. This is all part of a campus taking shape—but it is only preliminary to a college existing, because a college is more than a campus.

What would the observer have seen in these seven days of Genesis 1? At one level this could simply be dismissed as the wrong question. It continues to focus on the eyewitness account of material acts. But perhaps we can indulge our imagination for a moment as we return to the analogy of the college.

The main thing that happens is that students arrive. But even that would not necessarily mean much if faculty did not begin offering courses. In the light of those two events, however, everything else that was there all along takes on energy and meaning. The course schedule brings order to time. Time had been there all along, but the course schedule gives time a meaning to the college and the students. Even the course schedule had been there a long time (designed months earlier with students registering), but it has no existence until the semester begins. Dorms had existed filled with furniture. But now students inhabit the dorms and the furniture begins to serve its function.

The observer in Genesis 1 would see day by day that everything was ready to do for people what it had been designed to do. It would be like taking a campus tour just before students were ready to arrive to see all the preparations that had been made and how everything had been designed, organized and constructed to serve students. If Genesis 1 served as a liturgy to reenact (annually?) the inauguration of the cosmic temple, we also find a parallel in the college analogy as year by year students arrive and courses begin to bring life and meaning to the campus.

Death

Some might object that if the material phase had been carried out for long ages prior to the seven days of Genesis, there would be a problem about death. Romans 5:12 states unequivocally,

"Therefore, just as sin entered the world through one man, and death through sin, and in this way death came to all men, because all sinned." Interpreters have inferred from this verse that there was no death at any level prior to the Fall, the entrance of sin. But we should notice that the verse does not say that. Paul is talking about how death came to people—why all of humanity is subject to death. Just because death came to *us* because of sin, does not mean that death did not exist at any level prior to the Fall.

Not only does the verse not make a claim for death in general, everything we know logically repudiates the absence of death at any level prior to the Fall. Day three describes the process by which plants grow. The cycle of sprouting leaves, flowers, fruit and seeds is one that involves death at every stage. This system only functions with death as part of it. Likewise with animals: we need not even broach the topic of predatory meat eaters to see that the food chain involves death. A caterpillar eating a leaf brings death. A bird eating the caterpillar brings death. Fish eating insects brings death. If animals and insects did not die, they would overwhelm their environment and the ecology would suffer. Furthermore, if we move to the cellular level death is inevitable. Human skin has an outer layer of epidermis—dead cells—and we know that Adam had skin (Gen 2:23).

All of this indicates clearly that death did exist in the pre-Fall world—even though humans were not subject to it. But there is more. Human resistance to death was not the result of immortal bodies. The text indicates that we are formed from the dust of the earth, a statement of our mortality (for dust we are and to dust we shall return, cf. Gen 3:19). No, the reason we were not subject to death was because an antidote had been provided to our natural mortality through the mechanism of the tree of life in the garden. When God specified the punishment for disobedience, he said that when they ate, they would be doomed to death (the meaning of the Hebrew phrase in Gen 2:17). That punishment was carried out by banishing them from the garden and blocking access to the tree of life (Gen 3:23-24). Without access to the tree of life, humans were doomed to the natural mortality of their bodies and were therefore doomed to die. And so it was that death came through sin.

PROPOSITION II

"Functional Cosmic Temple" Offers Face-Value Exegesis

As discussed in chapter three when we explored the word *bārāʾ*, the word *literal* can have different meanings to different people. Mostly people use the word to express that they want to understand what the text "really says." The question is, what criteria make that determination? Certainly the meanings of words and the grammatical and syntactical framework are of importance. But grammar, words and sentences are all just the tools of communication. Usually our search to find out what a text "really says" must focus on the intended communication of the author and the ability of the audience to receive that same intended message. Words, grammar and syntax will be used adequately by a competent writer or speaker to achieve the desired act of communication. The same words can be used in a straightforward manner, or be used in a symbolic, metaphorical, sarcastic or allegorical way to achieve a variety of results.

As readers, we want to know how the author desired his communication to be understood. I referred to this in chapter three as the "face value" of the text. If a communication is intended to be metaphorical, the interpreter interested in the face value will want to recognize it as metaphor. If the author intends to give a history, the interpreter must be committed to reading it that way. In other words, interpreters have to give the communicator the benefit of the doubt and treat his communication with integrity.

Interpreters have come to Genesis 1 with a variety of approaches. Increasingly those who are uncomfortable with the scientific implications of the traditional interpretation have promoted a variety of ways to read the text so as to negate those implications. For example, some have suggested that the text is only theological—indicating that God is the Creator and the sabbath is important. Others have indicated that the text has a literary shape that makes it poetic and should not be taken as any sort of scientific record. While it is easy to affirm that important theology is the foundation of the account and that it has an easily recognizable literary shaping, one can still ask, is that all there is? Those who have championed the "literal" interpretation of the text have objected that these approaches are reductionistic attempts to bypass difficult scientific implications and claim that by pursuing them the text is so compromised that it is, in effect, rejected.

In the cosmic-temple interpretation offered in this book—which sees Genesis 1 as an account of functional origins—we find a different sort of resolution to the problems faced by the interpreter. I believe that if we are going to interpret the text according to its face value, we need to read it as the ancient author would have intended and as the ancient audience would have heard it. Though the literary form of expression and the theological foundation are undeniable, I believe that study of the ancient world indicates that far more is going on here than that.

Scholars in the past who have compared Genesis 1 to other ancient literature have sometimes suggested that the biblical text intends to be polemical—to offer a view in opposition to that of the rest of the ancient world. Again, it cannot be denied that Genesis offers a very different perspective than other creation texts in a number of ways. Here there is only one God, and there is no conflict to overcome. Since Genesis allows only one God, the account does not explain other gods being brought into existence and thus it breaks the close association between the components of the cosmos and the gods. All of this is true, and could be viewed as polemic. But it must also be noticed that the author of Genesis 1 is not explicitly arguing with the other views—he is simply offering his own view. His opposition to other ancient views is tacit.

The view presented in this book has emphasized the similarities between the ways the Israelites thought and the ideas reflected in the ancient world, rather than the differences (as emphasized in the polemical interpretation). While we can never achieve deep levels of understanding of how an ancient Israelite thought, we can at least see some of the ways they thought differently than we do. In this small accomplishment we can identify ways that we may have been inclined to innocently read our own thought patterns into texts whose authors did not share those thought patterns. If the Israelites, along with the rest of the ancient Near East, thought of existence and therefore creation in functional terms, and they saw a close relationship between cosmos and temple, then those are part of the face value of the text and we must include them in our interpretation.

In contrast, a concordist approach intentionally attempts to read an ancient text in modern terms. Concordist interpretations attempt to read details of physics, biology, geology and so on into the biblical text. This is a repudiation of reading the text at face value. Such interpretation does not represent in any way what the biblical author would have intended or what the audience would have understood. Instead it gives modern meaning to ancient words.

The rationale for this sort of reading involves several factors. First, these interpreters identify the ultimate author of Scripture as God. Therefore they feel justified in suggesting that reading the text scientifically yields God's intention even if the human author knew nothing of it. How do they determine the divine author's meaning if not through the human author? Their answer often derives from the idea that "all truth is God's truth." Therefore if we believe that physicists, biologists, geologists and other scientists have a bead on truth, that truth can be attributed to the divine author. Thus they might conclude that if the big bang really happened as a mechanism for the origins of the universe, it must be included in the biblical account of the origins of the universe. So concordists will attempt to determine where the big bang fits into the biblical record and what words could be understood to express it (even if in rather mystical or subtle ways). In this way the concordist is looking at modern science and trying to find a place for it in the biblical account with the idea that science has determined what really happened, so the Bible must reflect that. Other concordists rewrite science so that the correlation with the Bible can be made comfortably. In this way, concordism can be seen to be very different than wrestling with the face-value meaning of the text.

The problem with concordist approaches is that while they take the text seriously, they give no respect to the human author. The combination of "scientific truth" and "divine intention" is fragile, volatile and methodologically questionable. We are fully aware that what we call "scientific truth" one day may be different the next day. Divine intention must not be held hostage to the ebb and flow of scientific theory. Scientific theory cannot serve as the basis for determining divine intention.

God has communicated through human authors and through their intentions. The human author's communication is inspired and carries authority. It cannot be cast aside abruptly for modern thinking. The human author gives us access to the divine message. It has always been so. If additional divine meaning is intended, we must seek out another inspired voice to give us that additional divine meaning, and such an inspired voice can only be found in the Bible's authors. Scientific theory does not qualify as such an inspired voice.

We have neither the right nor the need to force the text to speak beyond its ken. This is not only important on a theoretical level, it is observable throughout the text. As mentioned in chapter one, there is not a single instance in the Old Testament of God giving scientific information that transcended the understanding of the Israelite audience. If he is consistently communicating to them in terms of their world and understanding, then why should we expect to find modern science woven between the lines? People who value the Bible do not need to make it "speak science" to salvage its truth claims or credibility.

The most respectful reading we can give to the text, the reading most faithful to the face value of the text—and the most "literal" understanding, if you will—is the one that comes from their world not ours. Consequently the strategy we have adopted for reading the text as ancient literature offers the most hope for treating the text with integrity. We are not trying to bypass what the text is saying, nor to read between the lines to draw a different meaning from it.

Concordist approaches, day-age readings, literary or theological interpretations all struggle with the same basic problem. They are still working with the premise that Genesis 1 is an account of material origins for an audience that has a material ontology. Modern inability to think in any other way has resulted in recourse to all of this variety of attempts to make the text tolerable in our scientific naturalism and materialism.

Our face-value reading in contrast, does the following:

1. recognizes Genesis 1 for the ancient document that it is;
2. finds no reason to impose a material ontology on the text;
3. finds no reason to require the finding of scientific information between the lines;
4. avoids reducing Genesis 1 to merely literary or theological expressions;
5. poses no conflict with scientific thinking to the extent that it recognizes that the text does not offer scientific explanations.

PROPOSITION 12

Other Theories of Genesis 1 Either Go Too Far or Not Far Enough

Previous chapters have made passing reference on a number of occasions to other theories concerning Genesis 1. In this chapter each one will be briefly evaluated to identify the points of comparison with the theory proposed here.

Young Earth Creationism (YEC)

The YEC position believes that the days in Genesis 1 are consecutive twenty-four-hour days during which the entire material cosmos was brought into existence. Proponents of this view therefore believe that everything must be recent (the origins of the universe, the earth and humankind). Some variation exists as to whether the cosmic origins go back 10,000-20,000 years as some would allow, or only go back about 6,000 years from the present (as promoted at the Creation Museum in Petersburg, Kentucky). The challenge they face is to account for all of the evidences of great age of the earth and of the universe. They do this by offering alternative theories allegedly based on science. For example, they typically account for the visibility of the stars by suggesting that light was created in transit. Most propose that the geological strata were laid down by the flood, and some contend that continental drift has all taken place since the flood. They commonly use the idea that God created with the appearance of age to account for some of what is observed.

Though each of their proposals could be discussed individually,[1] it is more important here to address the foundation of the approach. I would contend that this view goes too far in its understanding of what we need to do to defend the biblical text. It goes too far in its belief that the Bible must be read scientifically, and it goes too far in its attempts to provide an adequate alternative science. It uses a particular interpretation of the biblical text to provide the basis for scientific proposals about rock strata, an expanding universe and so forth. The YEC position begins with the assumption that Genesis 1 is an account of material origins and that to "create" something means to give it material shape. It would never occur to them that there are other alternatives and that in making this assumption they are departing from a face-value reading of the biblical text. In fact they pride themselves on reading the text literally and flash this as a badge of honor as they critique other views. Reading the text scientifically imposes modern thinking on an ancient text, an anachronism that by its very nature cannot possibly represent the ideas of the inspired human author.

I would contend that while their reading of the word "day" *(yôm)* as a twenty-four-hour day is accurate, they have been too narrow in their reading of words such as "create" *(bārāʾ)* and "made" *(ʿāśâ)*. It is not that they have considered the merits of a nonmaterial understanding of these words and rejected it. They are not even aware that this is a possibility and have therefore never considered it. In the functional view that has been presented in this book, the text can be taken at face value without necessitating all of the scientific gymnastics of YEC. Their scientific scenarios have proven extremely difficult for most scientifically trained people to accept. When the latter find YEC science untenable, they have too often concluded that the Bible must be rejected.

Old Earth Creationism (OEC)

One of the more prominent voices supporting the OEC position is found in the writings of Hugh Ross and his associates (Reasons to Believe). Ross believes that the Bible is not characterized by the limited scientific knowledge of its time and place.[2] So, for example, he suggests that in Genesis 1:3-5 the presence of light is evident through the "dense shroud of interplanetary dust and debris" that prevents the heavenly bodies from being seen. He sees day two as the beginning of the water cycle and "the formation of the troposphere, the atmospheric layer just above the ocean where clouds form and humidity resides, as distinct from the stratosphere, mesosphere, and ionosphere lying above." He looks to tectonics and volcanism to explain day three.[3] Ross believes, along with many others, that the old age of the earth and the universe can be easily accommodated to Genesis 1 once we realize that the days can represent long eras.[4]

One may not be inclined to dispute the science that underlies this approach, and Ross's desire to validate the text of Genesis, as in the YEC camp, is commendable. The question is, Is that what the author of Genesis is trying to say? We might be able to make the claim that there is some sort of compatibility between the scientific sequence and the textual sequence, but that is not proof that the text should be interpreted in scientific ways with advanced scientific content (latent in the text). One could do the same thing with Babylonian or Egyptian creation accounts. It is proof of our ingenuity rather than evidence of some ingrained underlying science.

If those from this camp were to consider the merits of the functional view proposed in this book, they would not have to give up all the scientific correlations proposed, but such an approach would no longer be of interest or carry any urgency, necessity or significance. They would only have to admit that the text makes no such claims and requires no such validation. Taking the text seriously is not expressed by correlating it with modern science; it is expressed by understanding it in its ancient context. If the text is interested in functional origins, it need not be evaluated against material claims and material knowledge. Its validation would come in answer to the question, Is this really how God set up the world to run, and is he the one who set it up? This stands in stark contrast to the validation that asks, Is this a scientifically accurate account of how the material universe came into being?

Framework Hypothesis

The framework hypothesis represents a literary/theological approach to Genesis 1. On the literary side it recognizes that the account of the seven days is highly structured, with the first three days defining realms of habitation and the second set of three filling these realms with inhabitants. Parallels exist between days one and four, days two and five, and days three and six. From this literary structuring conclusions are drawn about the account.

> We may simply conclude from this high level of patterning that the order of events and even lengths of time are not part of the author's focus. . . . In this understanding, the six workdays are a literary device to display the creation week as a careful and artful effort.[5]

Discussion then typically follows that draws out the theologically significant points of the passage on which all agree: God as Creator of all, the sovereignty of God, the power of the spoken word, the "goodness" of creation, the image of God in people and the significance of sabbath.[6]

The question to be posed to this group is whether they have gone far enough with the text. Is there more to it than theological affirmations expressed in a literary way? While no objection can be raised against the literary structure and no disagreement with the theological points, one has to ask whether Israelites thought of this text in only literary/theological terms. This view risks reductionism and oversimplification, and should be only a last resort.

For those who have in the past adopted the framework hypothesis, the theory proposed in this book does not require them to discard that interpretation, but only to accept the functional perspective alongside it. This does not require replacement, but would add value.

OTHER THEORIES

Throughout much of the twentieth century, a popular view was known as the "gap theory" or the "ruin-reconstruction" theory, promoted in the Scofield Reference Bible. It suggested that Genesis 1:1 recounted a prior creation ruled by an unfallen Satan. It had the advantage that it allowed for the universe and earth to be old, but the days of Genesis to be recent. Anything that did not fit into a recent earth (e.g., geological strata, dinosaurs) could just be shoved back into the first creation. In this view, at Satan's fall that first creation was destroyed—this is the gap between Genesis 1:1 and Genesis 1:2. The second verse was translated, "The earth became formless and void." Response to this theory demonstrated that the Hebrew text could not be read in that way and the theory has been gradually fading from the scene.

Others have suggested that the accounts in Genesis 1:1—2:3 and Genesis 2:4-25 are separated by many millions of years. In this view the old earth can be supported along with the mass appearance of hominid species in the first account. The second account is then associated with something like the Neolithic revolution in relatively recent times and associated with the granting of the image of God on two individuals that leads to Homo sapiens.[7] The problems with this position are largely theological. Were the previous hominid species in the image of God? Were they subject to death? How do they relate to the Fall? Are they biologically mixed into the current human race? These are questions that need to be answered by those promoting this position.

In conclusion it should be reemphasized that all of these positions have in common that they are struggling to reconcile the scientific findings about the material cosmos with the biblical record without compromising either. They all assume that the biblical account needs to be treated as an account of material origins, and therefore that the "different" scientific account of material origins poses a threat to the credibility of the biblical account that has to be resolved. This book has proposed, instead, that Genesis 1 was never intended to offer an account of material origins and that the original author and audience did not view it that way. In fact, the material cosmos was of little significance to them when it came to questions of origins. In this view, science cannot offer an unbiblical view of material origins, because there is no biblical view of material origins aside from the very general idea that whatever happened, whenever it happened, and however it happened, God did it.

Technical Support

Blocher, Henri. *In the Beginning.* Downers Grove, Ill.: InterVarsity Press, 1984.

Carlson, Richard. *Science and Christianity: Four Views.* Downers Grove, Ill.: InterVarsity Press, 2000.

Moreland, J. P., and John Mark Reynolds, eds. *Three Views on Creation and Evolution.* Grand Rapids: Zondervan, 1999.

Ratsch, Del. *The Battle of Beginnings.* Downers Grove, Ill.: InterVarsity Press, 1996.

PROPOSITION 13

The Difference Between Origin Accounts in Science and Scripture Is Metaphysical in Nature

We have now completed the presentation of the view that Genesis 1 presents an account of functional origins and will begin to integrate this view into the broader issues of science and society. The following chapters will explore the implications of this view in relation to evolution and Intelligent Design, as well as a consideration of some of the issues of policy in public education. As a prologue to that discussion, this chapter will draw some distinctions at the metaphysical level that will seek to probe some of the philosophical questions and reality outside of the material realm.

Many people who feel caught in a perceived origins conflict between the Bible and science subconsciously think of the origins question as a pie. Various aspects of origins are evaluated to decide whether God did it or a naturalistic process could be identified. The "origins pie" is then sliced up with each piece either going to "supernatural" or "natural" causation. The inevitable result as science progresses is that God's portion gets smaller and smaller, and overall, God becomes no longer useful or necessary.

Chapter one already discussed the issue that the distinction between "natural" and "supernatural" is not readily evident in the Old Testament and its world. One could go through passages such as Psalm 104 or Job 38 and see that the things attributed to God can also be explained in "natural" terms. The ancients were not inclined to distinguish between primary and secondary causation, and everything was attributed to deity. We can see, then, that the pie model is characterized by a distinction that is essentially unbiblical.[1]

If we want to adopt a more biblical view, we have to switch desserts! We need to think in terms of a layer cake.[2] In this view the realm of scientific investigation would be represented in the lower layer. This layer represents the whole realm of materialistic or naturalistic causation or processes. It is subject to scientific observation, investigation and explanation. Discovery in this layer does not subtract from God or his works. This is the layer in which science has chosen to operate and where it is most useful.

In contrast, the top layer represents the work of God. It covers the entire bottom layer because everything that science discovers is another step in understanding how God has worked or continues to work through the material world and its naturalistic processes. In this way, the bottom layer might be identified as the layer of secondary natural causation while the top layer is identified as ultimate divine causation.[3]

Science, by current definition, cannot explore the top layer. By definition it concerns itself with only that which is physical and material.[4] By restricting itself to those things that are demonstrable, and more importantly, those things that are falsifiable, science is removed from the realm of divine activity. Though scientists have their beliefs, those must be seen as distinct from their scientific work. It is unconvincing for a scientist to claim that he or she finds no empirical evidence of God. Science as currently defined and practiced is ill-equipped to find evidence of God.[5] The bottom layer may continue to have areas for which science cannot offer explanation, but that is only evidence of science's limitations, not evidence of God. A believer's faith holds that there is a top layer, even though science cannot explore it.

That top layer addresses ultimate causation, but it also addresses purpose, which in the end, is arguably more important. God is always the ultimate cause—that is our belief whatever secondary causes and processes can be identified through scientific investigation. But we also believe that God works with a purpose. Neither ultimate cause nor purpose can be proven or falsified by empirical science. Empirical science is not designed to be able to *define* or *detect* a purpose, though it may theoretically be able to deduce rationally that purpose is logically the best explanation.[6] As the result of an empirical discipline, biological evolution can acknowledge no purpose, but likewise it cannot contend that there is no purpose outside of a metaphysical conclusion that there is no God. It must remain neutral on that count since either contention requires moving to the top layer, which would mean leaving the realm of scientific inquiry. Science cannot offer access to God and can neither establish his existence beyond reasonable doubt nor falsify his existence. Therefore science can only deal with causation sequences—it cannot establish beyond reasonable doubt that a purpose governs or does not govern that which they observe.

The term for the technical philosophical interest in purpose is *teleology*. Teleology is the study of the goal of some intentional process that is usually the byproduct of purpose. That is, God works intentionally with his own purposes in mind to achieve a final goal. This concerns the realm of theology, or more broadly, metaphysics, and is not the stuff of empirical science.

The scientific observations and theories that compose the lower layer of the cake do not in and of themselves carry teleological conclusions (though they might be consistent with such conclusions). They cannot do so, because the presence of a purpose cannot be falsified. So some scientists might believe that the lower layer is all there is. For them the naturalistic causes are all that can be affirmed, and they do not believe in a purpose, for their layer, their worldview, their metaphysics, have no room for God. This view is exclusively materialistic and could be described as *dysteleological* (no discernible purpose).[7] This is not a scientifically drawn conclusion, but one that is drawn from the limitations of science. It would be like a fish claiming that there was only water, no air (despite the fact that they could not breathe if the water were not oxygenated by the air).

In contrast, there are many scientists who believe that there is indeed a top layer—that there is a God involved in ultimate causes and carrying out his purposes through the naturalistic operations of the cosmos. This belief does not change their approach to their scientific study—it does not affect their perception of the bottom layer nor does it affect their methods for studying the bottom layer. But their metaphysical position would be described as teleological. Nothing is random or accidental. Many of the great minds in the history of science were in this category (e.g., Galileo, Newton).

I have proposed here that Genesis is *not* metaphysically neutral—it mandates an affirmation of teleology (purpose), even as it leaves open the descriptive mechanism for material origins. Affirming purpose in one's belief about origins assures a proper role for God regardless of what descriptive mechanism one identifies for material origins. Since Genesis is thoroughly teleological, God's purpose and activity are not only most important in that account, they are almost the only object of interest. Genesis is a top-layer account—it is not interested in communicating the mechanisms (though it is important that they were decreed by the word of God). Whatever empirical science has to say about secondary causation offers only a bottom-layer account and therefore can hardly contradict the Bible's statements about ultimate causation. Whatever mechanisms can be demonstrated for the material phase, theological convictions insist that they comprise God's purposeful activity. It is not a scientific view of mechanism (naturalism) that is contrary to biblical thinking, but exclusive materialism that denies biblical teaching. Naturalism is no threat—but materialism and its determined dysteleology is.[8]

The functional orientation proposed for Genesis 1 in this book is fully in line with a penetrating teleology. God's purposes and intentions are most clearly seen in the way the cosmos runs rather than in its material structure or in the way that its material structures were formed (although the material structures can point to a designer). Instead of offering a statement of causes, Genesis 1 is offering a statement of how everything will work according to God's purposes. In that sense the text looks to the future (how this cosmos will function for human beings with God at its center) rather than to the past (how God brought material into being).[9] Purpose entails some level of causation (though it does not specify the level) and affirms sovereign control of the causation process.

The principle factor that differentiates a biblical view of origins from a modern scientific view of origins is that the biblical view is characterized by a pervasive teleology: God is the one responsible for creation in every respect. He has a purpose and a goal as he creates with intentionality. The mechanisms that he used to bring the cosmos into material existence are of little consequence as long as they are seen as the tools in his hands. Teleology is evident in and supported by the functional orientation.

TECHNICAL SUPPORT

Lamoureux, Denis. *Evolutionary Creation.* Eugene, Ore.: Wipf and Stock, 2008.

PROPOSITION 14

God's Roles as Creator and Sustainer Are Less Different Than We Have Thought

Now that we have developed a modified view of the creation account in Genesis and a corresponding modified view of what constitutes creative activity, we can explore how these give us a renewed vision of God as Creator.

Two extremes need to be avoided as we seek to understand God as Creator:

1. that his work as Creator is simply a finished act of the past (potential for deism), or
2. that his work as Creator is in an eternally repeating present (potential for micromanagement)

The first extreme is most common in popular Christianity today. In this view Genesis is an account of material origins and the creation of the physical universe took place in the past (whether the distant past or the more recent past). Consequently God's role as Creator was focused on a particular time and a particular task, and has been completed. This view can easily result in a practical deism in that it generally assumes that in creation God set up natural laws and physical structures subject to those laws so that the universe now virtually "runs by itself." This view potentially distances God from the day-to-day operations of the cosmos.

One form of this practical deism is particularly noticeable in some permutations of "theistic evolution" in which God is seen as responsible for "jump-starting" the evolutionary process and then letting it unwind through the eons. Alternatively God is sometimes viewed as involved more regularly at critical junctures to accomplish major jumps in evolution. The problem is that these approaches not only potentially remove God from ongoing operations in nature, but they even write God out of most of the origins story. The deism view gives too much to the ongoing functions of creation as well as rendering them too independent from God. The interventionist view treats the functionality of natural processes too lightly, as being inadequate to accomplish God's purposes. Potentially, the processes left to run on their own might very well fail to achieve God's purposes, but this possibility reveals the all-or-nothing assumption behind these two views—that what happens in natural history is either all due to natural processes running on their own *or* is due to direct divine intervention in the natural operations. That God might be working alongside or through physical and biological processes in a way that science cannot detect is one possibility that this either-or assumption ignores.[1]

But in all fairness the young-earth creationists are not immune from distancing God from the operations of nature. Even though they view God as totally responsible for origins, his Creator work is considered finished after those first six days. The "natural" world has been put in place, and it runs (on its own? vaguely sustained?) by those principles God put in place. For those who see it that way (admittedly not all in this camp), creation is over, and a practical deism looms over the ongoing operations of the world.

A second extreme, rather than adopting the sharp discontinuity between creation and operations as just described, considers there to be such continuity that it virtually eliminates beginning and end. Here creation is a constantly recurring process,[2] and God never ceases creating. One immediate objection to this view is found in the idea of teleology that was presented in the last chapter. For there to be a goal and purpose (telos), there must be a beginning and an end.[3] But beyond this important distinction, we need to explore the nature of continuity and discontinuity between the creative acts in Genesis 1 and what might be considered continuing creative activity.

The Bible to some extent offers the idea that creation is ongoing and dynamic. So theologian Jürgen Moltmann believes that God's creative work is not just the static work of the past, but that it is dynamic as it continues in the present and into the future.[4] This suggestion merits consideration, but key to the discussion is the extent to which what happens after the beginning could still be called creation, or if it is something else (e.g., "sustaining").[5] The answer to this question may be determined by how we understand the nature of creative activity in the Bible, and particularly, the view of origins underlying Genesis 1.

In the position of this book, the idea that Genesis 1 deals with functional origins opens up a new possibility for seeing both continuity and a dynamic aspect in God's work as Creator, because he continues to sustain the functions moment by moment (for example, see Neh 9:6; Job 9:4-10; Job 38; Ps 104; Ps 148; Amos 4:13; Mt 6:26-30; Acts 17:24-28; Col 1:16-17; Heb 1:3).[6] Creation language is used more in the Bible for God's sustaining work (i.e., his ongoing work as Creator) than it is for his originating work.[7] As we reduce the distinction between creating and sustaining, we take a departure from Moltmann, whose idea of dynamic creation considers all of covenant, redemption and eschatology as creative acts.[8]

I contend that there is a line between the seven days of Genesis 1 and the rest of history, making Genesis 1 a distinct beginning that is located in the past. If we see this as an account of functional origins, the line between is dotted rather than solid, as the narrative of Genesis 1 puts God in place to perpetuate the functions after they are established in the six days. In this way, day seven, God taking up his rest in the center of operations of the cosmos, positions him to run it. This continuing activity is not the *same* as the activity of the six days, but it is the reason why the six days took place. John Stek summarizes it well as he states that "in the speech of the Old Testament authors, whatever exists *now* and whatever *will come into existence* in the creaturely realm has been or will have been 'created' by God. He is not only the Creator of the original state of affairs but of all present and future realities."[9] As noted several times already, this does not result in a view of God as a micromanager, but it insists that he cannot be removed from the ongoing operations. The paradox of intimate involvement without micromanagement defies definition.

Returning to the college analogy that we introduced earlier, the origin of the college was intentional, with purpose in mind—all of the courses were designed, faculty and staff hired, students enrolled so that the college could exist. Those functions must continue to be sustained for the college to remain in existence, and it is the ongoing work to keep the college running that constitutes its dynamic aspect. Once the college (or cosmos) is brought into existence, that functional existence must be continually sustained. The physical campus must be *maintained* (cleaned, kept up, repaired, etc.), but the functional college must be *sustained* (courses offered again and again, new students enrolled, new employees hired, etc.). *Maintaining* relates to the material and the physical existence. *Sustaining* relates to the functional and operational. Consequently, when we take the functional approach to origins and the theological position of God's continual sustaining work, both originating and sustaining can be seen as variations of the work of the Creator, even though they do not entirely merge together. Genesis 1 is in the past, but the continuing activities of the Creator in the future and present are very much a continuation of that past work. In contrast to the first extreme, creation is not over and done with. In contrast to the second extreme, origins is rightfully distinguished from God's sustaining work, but both could be considered in the larger category of creation.

As we are going to discuss in the remainder of the book, it is precisely this pervasive role of God as Creator in all aspects of originating and sustaining that serves as the main dispute that Christians have with a purely materialistic view of origins. This materialistic view is often interwoven with biological evolution and at times is referred to as "evolutionism."[10] The existence of biological processes is not a major concern, whereas the denial of any role to God in relation to those biological processes—whatever they are—are theologically and biblically unacceptable. But that discussion is for another chapter.

The relationship between creation and other aspects of God's work such as covenant, redemption and eschatology is that each of these also involves God in the process of bringing order to disorder. He also did this for the cosmos in his creating work and continues to do it in sustaining the cosmos. But these—covenant-making, redemption and so on—are more related to his role in progressive revelation than to his Creator role.

In conclusion I suggest that God initiated the functions in Genesis 1 so that they are seen to originate in him. As a result of taking up his residence in the cosmic temple, he sustains the functions moment by moment, as the very existence of the cosmos depends on him entirely. Both initiating and sustaining are the acts of the Creator God. We recognize his role of Creator God by our observance of the sabbath, in which we consciously take our hands off the controls of our lives and recognize that he is in charge. His place in the temple and his role as Creator may have been ritually reenacted annually in temple liturgies. It would be a commendable sacred holiday for the church to reinstate. For even though God does not reside in geographical sacred space any longer, he is still in his cosmic temple, and he now resides in the temple that is his church (1 Cor 3:16; 6:19).

Technical Support

Fretheim, Terence E. *God and World in the Old Testament: A Relational Theology of Creation.* Nashville: Abingdon, 2005.

PROPOSITION 15

Current Debate About Intelligent Design Ultimately Concerns Purpose

HAVING NOW COVERED THE BIBLICAL and theological issues, we are ready to move into the discussion of contemporary issues. Specifically the next several chapters explore the impact of this view of Genesis 1 on our understanding of evolution, Intelligent Design and public education.

As we begin, it is most important to keep in mind that the view presented in the preceding chapters is what philosophers would label as "teleological"—by which they mean that the view involves God working with intention, purpose and a goal in every aspect of his role as Creator (which includes originating and sustaining). The obvious result of this is that all of creation is, by this definition, intelligent, and likewise, all of it is designed. Nothing could be considered accidental. Nothing happens "by itself," and origins are not just found in the outworking of natural laws. Nothing is really coincidence. In one of Orson Scott Card's novels one character quips, "Coincidence is just the word we use when we have not yet discovered the cause. . . . It's an illusion of the human mind, a way of saying, 'I don't know why this happened this way, and I have no intention of finding out.'"[1]

Likewise, the fact that we believe that God did *X* does not mean that it is no longer subject to scientific investigation. Everything that exists and everything that happens is, in Christian thinking, ultimately an act of God. Yet in the layer cake model we have presented, that does not mean that scientific or historical inquiry should be cut off—they still have the potential of leading to understanding at a different level.

In recent decades a movement referred to as Intelligent Design has become prominent. Throughout the ages scientists have always admired the cosmos as evidencing design, though in more modern times, many scientists are more likely to talk about the "appearance" of design. The Intelligent Design movement (ID) insists that this appearance of design is not illusive, but is the result of an unidentified intelligent designer.

One of the primary ways the Intelligent Design movement has offered evidence for its contention is through the identification of what they call irreducible complexity.[2] They have identified structures that require a multitude of parts that need to be functional all at once for the structure to continue to exist and do its job, therefore concluding that the structure could not have evolved one piece at a time. They make no consistent claims about the nature of the designer. They believe that these irreducible complexities show the weaknesses of Neo-Darwinian evolution (the reigning paradigm for understanding biological origins), but they have not gotten to the point where they have alternative scientific mechanisms to promote. In other words, ID does not offer a theory of origins. It offers conclusions from observations in the natural world and posits that those observations argue against the reigning paradigm of Neo-Darwinism. It must be noted, however, that even as many might grant weaknesses in the reigning paradigm, ID would only be one among many possible alternatives.

Protagonists of ID would like their claims and particularly their critique accepted as science. In the political realm, some have tried to force its adoption as an alternative to be offered in public education. The difficulty they face is that if there is intelligent design, there must logically be an intelligent designer. Given the existence of a designer, it would logically be inferred that such a designer is not simply playing games or being artistic, but is working with a purpose.[3] Science is not capable of exploring a designer or his purposes. It could theoretically investigate design but has chosen not to by the parameters it has set for itself (back to the layer cake analogy). Therefore, while alleged irreducible complexities and mathematical equations and probabilities can serve as a critique for the reigning paradigm, empirical science would not be able to embrace Intelligent Design because science has placed an intelligent designer outside of its parameters as subject to neither empirical verification nor falsification.

In short, teleological aspects (exploration of purpose) are not in the realm of science as it has been defined and therefore could not be factored into a scientific understanding. ID *could* be considered as contributing to the scientific enterprise when it is offering a critique of the reigning paradigm because it offers scientific observations in its support. But it does not contribute to the advance of scientific understanding because it does not offer an alternative that is scientifically testable and falsifiable. Its basic premise is a negative one: that "naturalistic mechanisms (i.e., natural selection, random mutation) cannot fully account for life as we know it."[4] ID does not deny the operation of naturalistic mechanisms—it simply finds them insufficient to offer a comprehensive explanation of all observable phenomena. It cannot offer at present a scientific hypothesis proposing alternatives. Consequently it can only offer inferences regarding science that can only be tracked currently by leaving the realm of science. Nevertheless proponents of ID would make a lesser claim that design itself is detectable and researchable and therefore can be subject to scientific investigation—the design element, not the nature or existence of the designer. They offer no theory of origins nor do they attempt to interpret the Bible or contribute to theological thinking.

Some would say that it is just plain and simple logic that some things are the product of design.

> Design seems to be a common thread that runs through the whole of nature. Time and again, in cases that have been cataloged since the dawn of biology, nature reveals that (1) its inhabitants are remarkably suited to fit their environment and (2) the various parts and systems that constitute organisms are remarkably suited to work in concert with one another.[5]

No one finds a watch on the beach and thinks that it is a relic of nature; no one looks at Mount Rushmore and concludes that it is the result of wind and erosion. But when these products of intelligent design are recognized, the process to understand them becomes a historical one, not a scientific one. To recognize them as products of design is to remove them from the arena of scientific investigation.

Intelligent Design has been criticized as being a God of the gaps approach. "God of the gaps" says that if there is no known naturalistic explanation of an observable phenomenon, that phenomenon is attributable to God. The unfortunate result of this way of thinking is that as scientific knowledge grows and more phenomena are explained, the role of God shrinks away. While ID vehemently denies being a God of the gaps approach, the logical hurdle is that if they believe that naturalistic explanations are insufficient, design in nature can only be established beyond reasonable doubt if all naturalistic explanations have been ruled out.[6]

Proving a negative logically requires that all possibilities have been considered, which in turn requires that all possibilities are known. As a result design cannot be established beyond reasonable doubt (it would be presumptuous to suggest that knowledge is so exhaustive that all possibilities are known), and it can only fall back on the claim that the *currently proposed* naturalistic mechanisms do not suffice. Design is thus attributed to observable phenomena that carry characteristic hallmarks of design (in an ID way of understanding) that cannot be explained by naturalistic mechanisms. This list ends up looking very much like the God of the gaps list.

Neo-Darwinism (N-D) is in no more attractive a position. While ID says that irreducible complexity provides evidence for design, N-D swings the pendulum in the opposite direction. It responds to the claims of irreducible complexity by proposing components that might have come together to produce what now appears to be irreducibly complex. Even if such an explanation cannot be found, or is criticized as being far-fetched, the underlying assumption is that there *must be one* (presumably because all phenomena *must* be the result of naturalistic mechanisms). Both then are ultimately based on metaphysical premises. ID has defined itself to allow a metaphysical acceptance of purpose (teleology), while the proponents of N-D presuppose by definition a metaphysical acceptance of "dysteleology"—that there can be no purpose or goal. In effect then ID suggests that there is warrant for opening scientific investigation to teleological possibilities. Mainstream science contends that dysteleology must be retained in its self-definition. At this point they are not willing to rewrite the current rules of science to allow for either intelligence or design. Having said this, it must be reiterated that whatever definitions of science may be and whatever scientific methods may be allowed or disallowed, the existence of purpose is unaffected.

Perhaps there are other naturalistic mechanisms beyond random mutation and natural selection that offer better explanations for observable phenomena (and along the way show more promise of explaining how presumably irreducibly complex phenomena came to be). Just such approaches are constantly being proposed and developed. What has been referred to as "meta-Darwinism"[7] includes a variety of (independent) proposals for naturalistic mechanisms that do not supplant natural selection and random mutation, but relegate them to a different role in the developmental process of organisms. These proposed mechanisms include endosymbiosis, developmental mutations (evo-devo), multilevel selection and complexity theory (self-organization). Of course these do not resolve the metaphysical issues if they still operate with dysteleological presuppositions. Some, to their credit, attempt to be neutral with respect to teleology. The stricture remains against making any explicit appeal to purpose in scientific explanations. To appeal to purpose is to shift to a different kind of explanation (e.g., metaphysical, theological).

Consequently we find that even as ID proposes that N-D fails to provide adequate naturalistic mechanisms to explain the existence of "irreducible complexities," the response of science has *not* been to admit that there must be a designer. Instead critique from a variety of sources has prompted continuing work to offer alternative naturalistic mechanisms that will remedy the inadequacies of N-D. This is how science works—it seeks out other scientific explanations. If scientists simply threw up their hands and admitted that a metaphysical, teleological explanation was necessary, they would be departing from that which is scientific.

The question is whether we can assume such hard and fast lines of distinction between the scientific and the metaphysical. It is true that observations can be put into one category or the other, but the fact is that such a categorization is artificial because none of us has a worldview comprised of only one of them. Science and metaphysics blend together in life. Can science be taught with no metaphysical aspect? Should metaphysics be isolated from the sciences? These questions will be dealt with in future chapters.

In conclusion, this chapter has introduced ID as both a critique of N-D, in which sense it alleges to be scientific, but also as offering an understanding of the world that is ultimately teleological—purposeful—in which sense it departs from the realm of scientific investigation and theorization.

The view of Genesis offered in this book is also teleological but accepts that all of creation is the result of God's handiwork, whether naturalistic mechanisms are identifiable or not, and whether evolutionary processes took place or not. God has designed all that there is and may have brought some of his designs into existence instantaneously, whereas others he may have chosen to bring into existence through long, complicated processes. Neither procedure would be any less an act of God.

Technical Support

Behe, Michael. *Darwin's Black Box.* New York: Simon & Schuster, 1996.

Dembski, William. *Intelligent Design.* Downers Grove, Ill.: InterVarsity Press, 1999.

Fowler, Thomas B., and Daniel Kuebler. *The Evolution Controversy: A Survey of Competing Theories.* Grand Rapids: Baker Academic, 2007.

House, H. Wayne, ed. *Intelligent Design 101.* Grand Rapids: Kregel, 2008.

Johnson, Philip. *Darwin on Trial.* Downers Grove, Ill.: InterVarsity Press, 1991.

O'Leary, Denyse. *By Design or By Chance?* Minneapolis: Augsburg, 2004.

PROPOSITION 16

Scientific Explanations of Origins Can Be Viewed in Light of Purpose, and If So, Are Unobjectionable

THE VIEW OFFERED OF GENESIS 1 recognizes that it was never intended to be an account of material origins. Rather it was intended as an account of functional origins in relation to people in the image of God viewing the cosmos as a temple. Though the Bible upholds the idea that God is *responsible* for all origins (functional, material or otherwise), if the Bible does not offer an *account* of material origins we are free to consider contemporary explanations of origins on their own merits, as long as God is seen as ultimately responsible. Therefore whatever explanation scientists may offer in their attempts to explain origins, we could theoretically adopt it as a description of God's handiwork. Scientific discussions of origins include a variety of different sciences including physics, geology, biochemistry and biology. As we consider these areas we might say that if there was a big bang (the current leading scientific explanation adopted by physicists and cosmologists), that is a description of how God's creation work was accomplished. If it turns out that some other explanation works better, God was at work through that. If the universe is expanding, God is at work. If geological strata were laid down eon by eon, God is at work. If various life forms developed over time, God is at work. Since biological evolution is the hot spot for controversy, we will focus our attention on that aspect of origins.

One possible objection is that too much in an evolutionary system is difficult to reconcile to the character of God. While it has been noted over the centuries that the cosmos is ideally suited for human habitation (anthropic principle), we also observe many disturbing features.[1] Survival of the fittest seems cruel. Pseudogenes seem useless and wasteful. Why were chromosomal aberrations not corrected instead of just being transmitted down the line?

In response to this objection, note that when Job believed that his understanding of the world and how it worked could be reduced to a single model (retribution principle: the righteous will prosper; the wicked will suffer), his suffering took him by surprise and was without explanation. How could such a thing happen? Why would God do this? The book is full of Job's demand for an explanation. When God finally appears he does not offer an explanation, but offers a new insight to Job. By confronting Job with the vast complexity of the world, God shows that simplistic models are an inadequate basis for understanding what he is doing in the world. We trust his wisdom rather than demanding explanations for all that we observe in the world around us and in our own lives. Scientific theories offer explanations concerning how the world, which we attribute to God's design, works. The objection to evolution raised above asks *why* God would do it that way. This is one of those "if I were God I would do it differently" (read, "better") kinds of arguments that humans presumptuously engage in. This is unhelpful in the same way as questioning God's justice with the implication that we could do it better. God did what he did, and we cannot second guess him.

This is a lesson we still need to learn. God in his wisdom has done things in the way that he has. We cannot stand in judgment of that, and we cannot expect to understand it all. We can still explore the *what* and the *how* questions, but the *why* will always lie beyond our understanding and beyond our models. Relative to God, as humans we are by definition simplistic. We must also remember some of the key lessons of Scripture. In our weakness he is strong. He can use suffering to strengthen our character. He can use evil to accomplish good (precisely the nature of the discussion in the book of Habakkuk). God's sovereignty is demonstrated in that whatever personal or nonpersonal agents do, God takes it and turns it to his purpose.

Our question then cannot be whether one model or explanation for the cosmos and its origins is reconcilable with the nature of God. We don't have enough information to make that assessment. We can only ask what Scripture requires us to defend.

In chapter one we pointed out that the common dichotomy drawn today between "natural" and "supernatural" did not exist in the ancient world. I would also propose that it is not theologically sound. God cannot be removed or distanced from those occurrences that we so glibly label "natural." When we so label phenomena, it is an indication that we understand (at least to some extent) the laws and causes that explain it. Be that as it may, that does not mean that God does not control that process. What we identify as natural laws only take on their law-like quality because God acts so consistently in the operations of the cosmos. He has made the cosmos intelligible and has given us minds that can penetrate some of its mysteries.

Let us take an example to comment on this dynamic. In Psalm 139:13 the psalmist declares to God: "You knit me together in my mother's womb." This and other statements in the

Bible affirm God as the creator of each human being in the womb. The first observation is that this act of creation is not instantaneous but involves a process. Yet it is the work of God. A second observation is that this process is well understood by science. From the process of fertilization, implantation, fetal development and birth, scientists find that which is explainable, predictable and regular. The field of science called embryology offers a complex sequence of naturalistic cause and effect for the development of a child. Yet this blossoming of a life remains full of mystery.

Our biblical belief does not associate God's work only with those aspects that remain a mystery. God is involved with the entire process start to finish. He made us so that the process can work the way that it does, and each child is his handiwork.[2] In like manner we should observe that our biblical faith in the statement of Psalm 139 does not require us to denounce the science of embryology. It is not an either/or decision. God knits us together in our mother's womb and the processes observed by scientists merely explore the work of God. We have no cause to reject the science, yet science is incapable of affirming or identifying the role of God.

These same phenomena are also true in history. We believe that God is in control of history and shapes events moment by moment. It is all subject to his sovereignty. Despite that theological affirmation, no historian is able to see God's hand clearly, though depending on one's presuppositions one may conclude that God is at work. Some of those conclusions would be the result of incredible coincidences, while others would be the result of that which is otherwise unexplainable. We might notice that these are the same issues that drive Intelligent Design in their assessment of the sciences.

We believe that God controls history, but we do not object

when historians talk about a natural cause-and-effect process. We believe that God creates each human in the womb, but we do not object when embryologists offer a natural cause-and-effect process. We believe that God controls the weather, yet we do not denounce meteorologists who produce their weather maps day to day based on the predictability of natural cause-and-effect processes. Can evolution be thought of in similar terms?

It would be unacceptable to adopt an evolutionary view as a process without God. But it would likewise be unacceptable to adopt history, embryology or meteorology as processes without God. The fact that embryology or meteorology do not identify God's role, or that many embryologists or meteorologists do not believe God has a role makes no difference. We can accept the results of embryology and meteorology (regardless of the beliefs of the scientists) as processes that we believe describe in part God's way of working. We don't organize campaigns to force academic institutions that train meteorologists or embryologists to offer the theological alternative of God's role. Why should our response to evolution be any different?

There are, of course, some differences that come to mind. First, meteorology and embryology are advanced sciences—they are not taught in middle school. Therefore evolution is more of an issue in public education than the others are. Second, there is a sense in which evolution is "closer to home" in that it potentially touches on our identity, our place in the world, our sense of significance. As such it threatens us at personal levels in ways that meteorology and embryology do not. Third, the teaching of evolution is more likely to eventuate in metaphysical implications if not in explicit metaphysical statements. That is, it is more likely that evolution will be offered as an account of origins that explicitly denies God a role, thus setting up a conflict and demanding a choice. Such a choice is unnecessary and unacceptable (to be discussed in a future chapter), but should lead to adjustments in how the subject is taught, not in the total rejection of the principles and role of biological evolution.

This does not mean that all aspects of evolutionary theory should be accepted uncritically or even that evolution provides the best model. Meteorology and embryology are being constantly modified, and biological evolution is no different. I am not suggesting a wholesale adoption of evolution, merely suggesting that neither Genesis 1 specifically nor biblical theology in general give us any reason to reject it as a model as long as we see God as involved at every level and remain aware of our theological convictions.

As I have thought about the issues, it seems that there are three major reasons that people who take the Bible seriously have troubles with biological evolution.[3]

1. Theology

The problem people have on the theological level, as we have discussed, is that evolution is often construed in such a way as to leave God out of the picture—as if it denies the existence of God or even can establish beyond reasonable doubt that he does not exist. This is not a problem with evolutionary theory, only a problem with some who propagate evolution in dysteleological ways (absent of purpose). This problem is easily resolved by an affirmation that whatever evolutionary processes may have taken place, we believe that God was intimately involved in them. This is a metaphysical and theological decision that can only take place outside of the scientific aspects of evolutionary theory. The choice we make about God's role eliminates the problem without requiring that all evolutionary theory be rejected.

2. Genesis 1

Genesis 1 presents many challenges in people's minds to accepting evolutionary theory. As we have been discussing, many believe that the seven-day structure of Genesis 1 requires a young earth, while evolutionary theory requires long periods of time. Likewise some would point out that in Genesis 1 creation takes place by the word of the Lord, from which they infer instantaneous creation. The first of these objections is resolved if we see Genesis 1 to be an account of functional origins as proposed and defended in previous chapters. The question of the age of the earth can only be addressed from Genesis 1 if it is an account of material origins. If it is not, then the Bible offers no information on the age of the earth.

The second objection can be addressed by looking at the wide range of phenomena that are brought into being by divine speech (divine fiat). God is sovereign and his word is an effective decree. While some of what he decrees comes about immediately, in other instances his decree initiates a process.[4] One need not conclude that divine fiat implies instantaneous fulfillment. God does everything, and everything that he does is by his decree.

If Genesis 1 does not require a young earth and if divine fiat does not preclude a long process, then Genesis 1 offers no objections to biological evolution. Biological evolution is capable of giving us insight into God's creative work.

3. Genesis 2 and Romans 5

The third reason that people who take the Bible seriously object to evolution is related to the nature of humanity as being in the image of God, to the nature of sin, and to the question of the historicity of Adam and Eve. Here we are talking about theological realities taught clearly in the Old and New Testaments. How can human beings be considered the result of an evolutionary process and the biblical teachings be preserved? A solution that some offer suggests separating the material issues in human origins from the spiritual or metaphysical ones. In other words, they propose considering that humans develop physically through a process and somewhere in that process, undetectable by science, the image of God becomes part of the human being by an act of God. This would be followed by an act of disobedience by those image-bearing humans that constitutes the Fall and initiates the sin nature. Some suggest that this is what occurred with a single, historical human pair (a literal Adam and Eve) while others conjecture that this transpired with a group of persons so that "Adam and Eve" would be understood corporately as the first humans, not as a single original human pair. Such views, which I continue to find problematic on a number of levels, have been proposed in attempts to reconcile the supposed contradictions between the Bible and the anthropological fossil evidence, and they stand as examples of continuing attempts to try to sort out this complex issue.[5] Unfortunately no option is without difficulties.

As always, in our commitment to defend an accurate interpretation of the text and sound theology we must consider carefully and try to determine precisely what issues we must defend. The image of God and the sinful act of disobedience dooming all of humanity are biblical and theological realities linking us to Adam and Eve, whom the biblical text treats as historic individuals (as indicated by their role in genealogies).[6] That God is the Creator of human beings must be taken seriously. We continually seek understanding of biblical texts for what they communicate in their own theological and cultural contexts. Whatever evolutionary processes led to the development of animal life, primates and even prehuman hominids, my theological convictions lead me to posit substantive discontinuity between that process and the creation of the historical Adam and Eve. Rather than cause-and-effect continuity, there is material and spiritual discontinuity, though it remains difficult to articulate how God accomplished this. The point I want to make is that perhaps Genesis 2 and Romans 5 do not pose as many problems as some have thought, allowing us to reap from science understandings of how life developed up to and including the creation of the first humans.

If the theory proposed in this book is on target, Genesis 1 does not offer a descriptive model for material origins. In the absence of such a model, Christians would be free to believe whatever descriptive model for origins makes the most sense. The major limitation is that any view eventually has to give God full control of the mechanisms if it claims to be biblical. A biblical view of God's role as Creator in the world does not require a mutually exclusive dichotomy between "natural" and "supernatural," though the reigning paradigms are built on that dichotomy. It does not matter that there may be perfectly acceptable and definable empirical descriptions and explanations for observed phenomena and aspects of origins. Such would not exclude divine activity because without the natural/supernatural dichotomy, divine activity is not ruled out by empirical explanation. I can affirm with the psalmist that God "knit me together in my mother's womb" without denying the premises of embryology. Likewise those aspects of evolutionary mechanisms that hold up under scrutiny could be theoretically adopted as God's mechanisms.

TECHNICAL SUPPORT

Bube, Richard. *Putting It All Together.* Lanham, Md.: University Press of America, 1995.

Collins, C. John. *Science and Faith: Friends or Foes?* Wheaton, Ill.: Crossway, 2003.

Giberson, Karl. *Saving Darwin: How to Be a Christian and Believe in Evolution.* New York: HarperOne, 2008.

Glover, Gordon. *Beyond the Firmament.* Chesapeake, Va.: Watertree Press, 2007.

Hayward, Alan. *Creation and Evolution.* Minneapolis: Bethany House, 1985.

Lamoureux, Denis. *Evolutionary Creation.* Eugene, Ore.: Wipf and Stock, 2008.

Van Till, H. J., et al. *Science Held Hostage.* Downers Grove, Ill.: InterVarsity Press, 1988.

PROPOSITION 17

Resulting Theology in This View of Genesis 1 Is Stronger, Not Weaker

HAVING DISCUSSED WHAT EFFECT this interpretation of Genesis 1 has on thinking about science, we now ought to consider what effect it has on our thinking about theology. What threats might it pose or what strength or clarity might it offer?

The changes that this interpretation might suggest do nothing to weaken the picture of God. Even if the account in Genesis 1 is taken as an account of functional origins, it would not therefore imply that God is not responsible for material origins. The biblical view is that whatever exists from any perspective is the work of God. So this view does not reduce what God has done, it only suggests a change of focus concerning what aspect of God's work is represented in Genesis 1.

In the same fashion the suggestion that some of God's work of creation may have taken place over a long period of time rather than instantaneously does not reduce God's power.[1] God can create any way he sees fit, and it is no less an act of his sovereign power if he chooses to do it over extended billions of years. It is still accomplished by his word. Some would see the great span of time as further indication of God's majesty. If nothing is taken away from God's works and his sovereignty is not reduced, then there is no theological threat regarding God's person or deeds.

On the other side of the equation, there is much to be gained theologically from this interpretation of Genesis 1. In fact we will find that a more vital and robust theology of God as Creator emerges when we adopt this interpretation and its implications. Some of these have been pointed out in previous chapters, but here they will be gathered together for consideration.

God's Role in Everything

Our scientific worldview has gradually worked God out of the practical ways in which we think about our world. When science can offer explanation for so much of what we see and experience, it is easy for our awareness of God's role to drift to the periphery. It is not that we believe any less that he is active, it is just that we are not as conscious of his role. The result is a practical (if not philosophical) deism in which God is removed from the arena of operations.

In contrast, when God's work is fully integrated with our scientific worldview and science is seen to give definition to what God is doing and how he is doing it, we regain a more biblical perspective of the work—a perspective that is theologically healthier.

Creator Role Ongoing

If God's work of creation is considered only a historical act that took place in the past, it is easy to imagine how people might not think in terms of God being active today. We have lost the view that nature does not operate independently from God. He is still creating with each baby that is born, with each plant that grows, with each cell that divides, with each nebula that forms. We might find it easy to look at some majestic view like a glorious sunset or the grandeur of the mountains and ponder the magnificence of

God's handiwork. But this sense needs to extend beyond the "wow" moments to encompass all of our experience of his world. We have the same problem when we only recognize God in some incredible occurrence in our lives and forget that he provides for us, cares for us and protects us moment by moment, day after day. God did not just create at some time in the past; he is the Creator—past, present and future.

God's Control of Functions

Although we are acutely aware of the physical world around us, we live in a world of functions. Materialism sees the functions of our world as the consequence of structures, that is, that objects or phenomena in our world function the way that they do because of their physical structures. In the biblical way of thinking, the objects and phenomena in the world function the way they do because of God's creative purposes. This gets back to the issue of teleology that we have discussed in previous chapters. Materialism has no room for purpose, and so the operative equation concerns only structures and the resulting functions. The biblical way of thinking counters materialism when it insists that the most important part of the equation is God's purposes.

Our world tends to subordinate the functional to the material. That is why ever since the Enlightenment (at least) we have generally believed that it is most important for us to think of creation in terms of the material. Our world has taught us to give priority to the material. In the view that we have presented of Genesis 1, the material is subordinate to the functional. The Bible considers it much more important to say that God has made everything *work* rather than being content to say that God made the physical stuff. The purpose, the teleology (which is the most important part), is located and observed in the functional, not the material.

To think about the contrast between the material and the

functional, and the illusionary nature of the material world, consider the following statements of one of the characters in Orson Scott Card's novel *Prentice Alvin:*

> "Everything's mostly empty. That anvil, it looks solid, don't it? But I tell you it's mostly empty. Just little bits of ironstuff, hanging a certain distance from each other, all patterned there. But most of the anvil is the empty space between. Don't you see? Those bits are acting just like the atoms I'm talking about. So let's say the anvil is like a mountain, only when you get real close you see it's made of gravel. And then when you pick up the gravel, it crumbles in your hand, and you see it's made of dust. And if you could pick up a single fleck of dust you'd see that it was just like the mountain, made of even tinier gravel all over again."
>
> "You're saying that what we see as solid objects are really nothing but illusion. Little nothings making tiny spheres that are put together to make your bits, and pieces made from bits, and the anvil made from pieces—"
>
> "Everything is made out of living atoms, all obeying the commands that God gave them. And just following those commands, why, some of them get turned into light and heat, and some of them become iron, and some water, and some air, and some of them our own skin and bones. All those things are real—and so those atoms are real."[2]

Sacred Space

Once we turn our thinking away from "natural world" to "cosmic temple" our perspective about the world around us is revolutionized. It is difficult to think of the "natural world" as sacred (because we just designated it "natural"). When the cosmos is viewed in secular terms, it is hard to persuade people to respect it unless they can be convinced that it is in their own best interests to do so. If it is secular, it is easy to think of it only as a resource to be exploited. We even refer to "natural resources."

But when we adopt the biblical perspective of the cosmic temple, it is no longer possible to look at the world (or space) in secular terms. It is not ours to exploit. We do not have natural resources, we have sacred resources. Obviously this view is far removed from a view that sees nature as divine: As sacred space the cosmos is *his place*. It is therefore not *his person*. The cosmos is his place, and our privileged place in it is his gift to us. The blessing he granted was that he gave us the permission and the ability to subdue and rule. We are stewards.

At the same time we recognize that the most important feature of sacred space is found in what it is by definition: the place of God's presence. The cosmic-temple idea recognizes that God is here and that all of this is his. It is this theology that becomes the basis for our respect of our world and the ecological sensitivity that we ought to nurture.

SABBATH

The fourth commandment directs people to observe the sabbath based on God's rest in Genesis 1.[3] Throughout human history interpreters of Scripture have struggled to work out the implications of this directive. What constitutes rest? What activities are ruled out? Part of the difficulty is that the Bible offers little detail as it tends more toward vague generalizations. Furthermore most of the statements are negative (what one should not do) rather than positive (approved or even mandated activities).

Given the view of Genesis 1 presented in this book, we get a new way to think about the sabbath. If God's rest on the seventh day involved him taking up his presence in his cosmic temple which has been ordered and made functional so that he is now ready to run the cosmos, our sabbath rest can be seen in a different light. Obviously, God is not asking us to imitate his sabbath rest by taking the functional controls. I would suggest that instead he is asking us to recognize that he is at the controls, not us. When we "rest" on the sabbath, we recognize him as the author of order and the one who brings rest (stability) to our lives and world. We take our hands off the controls of our lives and acknowledge him as the one who is in control. Most importantly this calls on us to step back from our workaday world—those means by which we try to provide for ourselves and gain control of our circumstances. Sabbath is for recognizing that it is God who provides for us and who is the master of our lives and our world. We are not imitating him in sabbath observance, we are acknowledging him in tangible ways.

If we have to be reminded or coerced to observe it, it ceases to serve its function. Sabbath isn't the sort of thing that should have to be regulated by rules. It is the way that we acknowledge that God is on the throne, that this world is his world, that our time is his gift to us. It is "big picture time." And the big picture is not me, my family, my country, my world, or even the history of my world. The big picture is God. If the sabbath has its total focus in recognition of God, it would detract considerably if he had to tell us what to do. Be creative! Do whatever will reflect your love, appreciation, respect and awe of the God of all the cosmos. (This is the thrust of Is 58:13-14.) Worship is a great idea, but it can't be mechanical, and it may only be the beginning. It is up to the individual to determine his or her personal response to give the honor that is due. The more gratitude we feel toward God and the more we desire to honor him, the more the ceremonies will mean and the more we will seek out ways to observe the sabbath. All of this derives from a renewed understanding of the sabbath that proceeds from our interpretation of Genesis 1.

Order

Any reader of the Bible can see that wisdom is a worthy pursuit and that as an attribute of God he grants it to humans who, being in his image, are able to achieve it to some degree. What is less transparent, and often the topic of discussion, is exactly what constitutes wisdom. A theory I find very attractive for the way it suits the wide variety of data is that wisdom entails finding inherent order and conforming oneself to that order. One understands authority, society, family, relationships, ethics and etiquette all in relationship to an understanding of order.

Interpreters of Wisdom literature have consistently noticed how prominent a topic creation is in that literature. The connection of wisdom with order offers an explanation for that prominence. God's creative work has established order in the cosmos just as he has established order in society and all other areas. Science has observed that order and given us an appreciation of how deeply order penetrates.

In the interpretation of Genesis 1 that has been proposed here, we understand that one of the main emphases of the account of creation is the order that God brings to the cosmos in his wisdom. The temple was seen as being at the center of the ordered world as God established and preserved order in the world from the temple.

When we are troubled by the disorder that we encounter in this world, it is important to understand that the disorder and brokenness of this world are the result of human sin and the Fall. The theological commitment we draw from Genesis 1 is that God is the author of order. We respond by understanding how he has ordered the world: materially, functionally and spiritually.

Human Role

The description of humankind and the statement of blessing in

Genesis 1 can now be understood perhaps a little more clearly as related to human functions. When God grants the privilege that people may be fruitful and multiply, he gives us the function of populating the world without limitation. When God creates people in his image it indicates, perhaps among other things, that we are to function as his stewards over creation. When God gives the mandate to subdue and rule, he is assigning a task and providing the wherewithal to accomplish that task. Through Genesis 1 we come to understand that God has given us a privileged role in the functioning of his cosmic temple. He has tailored the world to our needs, not to his (for he has no needs). It is his place, but it is designed for us and we are in relationship with him.

This view is different from both the ancient Near East and different from modern materialism. In the ancient Near East people were created as slaves to the gods. The world was created by the gods for the gods, and people met the needs of the gods. In the Bible God has no needs, and his cosmic temple has been created for people whom he desires to be in relationship with him. In modern materialism people are nothing but physical forms having no function other than to survive. The theology of Genesis 1 is crucial to a right understanding of our identity and our place in the world.

Theological Implications of "It Was Good"

Finally, interpreters have often offered a variety of opinions of the meaning of the repeated statement in Genesis 1 that "it was good." Some have drawn far-reaching implications from their interpretation. We have already discussed in chapter four the idea that "good" is a reference to being functional, not a matter of moral goodness. This is an important distinction because it does not suggest that we ought to look for moral goodness in the way that the cosmos operates. When we think of "good" in connection to being functional rather than moral, we don't have to explain how predation can be part of a morally good world. As God indicated to Job, even though the world is God's place and functions under his control, that does not mean that the cosmos is a reflection of God's attributes (Job 38). The cosmos declares God's glory, and his existence can be deduced in the observation of the world, but those truths do not indicate that his attributes are consistently worked out in what we call the "natural world." Gravity is not just; rain falls on the righteous and unrighteous alike , even where no one lives (Job 38:25-27); the created world is not "fair." If it were going to be consistently fair and just, there would be no room for sin at all. Given that it is a sinful world, God's condescending grace reigns.

The theological issues presented in this list should be recognized as mirroring the theological interests about creation found in the rest of the Bible. As the reader of the Bible looks through Psalms, Wisdom literature, prophets and on into the New Testament, one finds these same sorts of theological affirmations to be the focus. The Bible gives little attention to material origins, though of course God did that too. Consequently even if the reader is not inclined to adopt the proposed interpretation of Genesis 1, his or her theology could still be greatly enhanced by the observations offered here by embracing a renewed and informed commitment to God's intimate involvement in the operation of the cosmos from its incipience and into eternity. We all need to strengthen our theology of creation and Creator whatever our view of the Genesis account of origins. Even though it is natural for us to defend our exegesis, it is arguably even more important to defend our theology. I have attempted to demonstrate that exegesis of the original meaning of Genesis 1 gives us no cause to argue with the idea of the physical world coming about by a slow process. But we do need to defend at all costs an accurate view of the nature of God and his role in our world.

So what affirmations does the proposed interpretation of Genesis 1 expect of us?

1. The world operates by Yahweh's design and under his supervision to accomplish his purposes.
2. The cosmos is his temple.
3. Everything in the cosmos was given its role and function by God.
4. Everything in the cosmos functions on behalf of people who are in his image.

PROPOSITION 18

Public Science Education Should Be Neutral Regarding Purpose

On the basis of the view that Genesis 1 is a discussion of functional origins, we may now tackle the question of what is appropriate in the classroom. If a science course intends to discuss material origins from the perspective of a material ontology (which is essential to the nature of empirical science), there is no point at which the Genesis account becomes relevant, because Genesis does not concern material origins and does not have a material ontology. A significant point of disagreement, however, does exist between the Bible and the metaphysical assumptions that may at times accompany the teaching of evolutionary theory. This conflict arises from the metaphysical issue of purpose (teleology). Framing the issue this way moves the discussion from the sphere of theology to the larger metaphysical sphere and asks: Are origins teleological (having a purpose and a goal) or dysteleological (no purpose, no goal)?

Those who accept the Bible by faith accept also by faith a teleological view of origins. Empirical science[1] is not designed to be able to *define* purpose, though it may theoretically be able to deduce rationally that purpose is logically the best explanation. As the result of an empirical discipline, biological evolution can acknowledge no purpose, but likewise it cannot contend that there is no purpose—it must remain teleologically neutral. In this book I have proposed that Genesis 1 presents an account of functional origins and therefore that it offers no descriptive mechanism for material origins. If this is so, one could accept biological evolution as providing a descriptive mechanism putatively describing how God carried out his purposes. Perhaps this approach could be labeled *teleological evolution*. In terms of cosmic origins, biblical theology is compatible with a descriptive mechanism such as that provided by biological evolution offered in terms that leave aside questions concerning purpose (i.e., teleologically neutral). But biblical theology is irreconcilable with metaphysical naturalism[2] to the extent that the latter is committed to refusing any consideration of purpose (dysteleological). This bone of contention concerns metaphysics, not empirical science.

I have proposed here that Genesis is *not* metaphysically neutral—it mandates an affirmation of purpose, but it leaves the descriptive mechanism for material origins undetermined. Teleological affirmation (there *is* a purpose and God is carrying it out in his work of creation) in one's belief assures a proper role for God regardless of the descriptive mechanism identified for material origins. This view of Genesis can be compared to other theoretical approaches as follows:

Creationism, particularly young earth creationism, differs from the view proposed in this book by insisting that the Bible *does* offer a descriptive mechanism for material origins in Genesis 1, and therefore is both teleological and intrinsically opposed to the descriptive mechanism offered by biological evolution. We have suggested that this perspective does not represent an accurate contextual reading of Genesis.

Biological evolution is an empirically derived model that suggests several descriptive mechanisms for material origins. As an empirically derived model, it can only be agnostic concerning teleological affirmation or denial because purpose cannot be identified by any empirical methods. The descriptive mechanisms associated with biological evolution *can* operate within empirical science without dabbling in the metaphysics of teleology. Of course that does not mean that this is how it is consistently handled in textbooks and classrooms. For example, in 1995 the National Association of Biology Teachers (NABT) issued a "Statement on the Teaching of Evolution."[3] An initial description of evolution used adjectives such as "unsupervised" and "impersonal."[4] These words faced strong opposition from a variety of outside parties and were later struck from the statement (1997 revision). More care is needed to articulate a view that, while unapologetic in its foundation in *methodological* naturalism,[5] avoids embracing *metaphysical* naturalism. A good example is in the revised statement that subsequently appeared on the NABT website, which indicates that "natural selection has no discernable [*sic*] direction or goal, including survival of a species."[6] The critical word here is *discernible,* which makes this a more carefully nuanced and more acceptable statement of metaphysical neutrality.

On the whole science educators seem very concerned about limiting their focus to that which is valid science.[7] This is commendable. So, for example, the NABT statements regularly have something like the following: "Evolutionary theory, indeed all of science, is necessarily silent on religion and neither refutes nor supports the existence of a deity or deities."[8] Unfortunately, although they are quick to dismiss positions which blatantly promote teleological perspectives (creationism, Intelligent Design), there seems to be no attempt to dismiss positions that blatantly promote dysteleology, which is equally impossible to affirm through empirical science (as they indicate). One wonders how willing the NABT would be to rise up against a teacher who actively promotes dysteleology, and would they do so with the same passion that they demonstrate when opposing those who support creationism or ID? Dysteleological approaches are just as invalid as teleological approaches in any curriculum that seeks to focus on empirical science. Science education can promote *methodological* naturalism (refusing to resort to a "God did it" explanation in their empirical study) without indoctrinating students in *metaphysical* naturalism, to which we now turn.

Metaphysical naturalism is *not* metaphysically neutral regarding teleology. Not content with an empirically based methodology, it mandates the restriction of *reality* to that which is material. By definition, empirical science is characterized by *methodological* naturalism, but once it begins propounding *metaphysical* naturalism, it has overstepped its disciplinary boundaries. We noted that Genesis assumes teleology (origins are the result of God acting with a purpose and a goal) and teaches tel-eology. That is part of its theology and is admittedly not something subject to observation or scientific demonstration—it is a matter of belief. Many modern scientists, in contrast, assume dysteleology (no purpose or goal), but such a conclusion is likewise part of a metaphysical system and is not subject to observation or scientific demonstration. Even when a divine hand cannot be observed through scientific methods, that is insufficient reason to conclude that a divine hand does not exist or is not active. Science is designed only to operate within the closed system of the material universe—it ought not therefore pass judgment on whether or not there is anything outside the material universe. It therefore should not draw dysteleological conclusions if it is seeking to restrict itself to valid science. This is an important observation in the discussion of public education.

Intelligent Design has been a subject of considerable controversy in recent years in the debate concerning public education. In our chapter on ID we drew a distinction between the issues of design and irreducible complexity, the former being largely metaphysical (though at times only reflecting a rational deduction), the latter reflecting a scientific observation about the interdependence of the parts of a structure. "Design" implies an intelligent cause rather than an undirected process, and as such proposes a *solution* to some perceived problems in biological evolution. The problem is that design refers to a rational deduction and as such is only one possible inference from what appears to some to be irreducible complexity.[9] Design by its nature can hardly avoid the transition from a rational deduction to a metaphysical proposal accompanied by an assumption of purpose (thus affirming a teleological view).[10] In contrast we have observed that evidence or claims of irreducible complexity can offer challenges for standard biological evolutionary theory. Such evidence confronts the reigning paradigm by raising questions about theories of evolutionary mechanisms that beg for solutions.

If public education is committed to the idea that science courses should reflect only empirical science, neither design nor metaphysical naturalism is acceptable because they both import conclusions about purpose into the discussion. This is not an issue of God, religion, faith, or church and state. It is a question about whether the metaphysical questions about purpose (teleology) should come into play in the science classroom, presumably adulterating that which is empirical with that which is nonempirical; and we contend that it should not.[11] The assertions of purposelessness (dysteleology) by materialists are objectionable to many people of faith, and affirmation of purpose (teleological elements) of theism, creationism or design are objectionable to many scientists. Once we rule out those approaches that represent blatant and self-acknowledged teleological platforms (i.e., Genesis, creationism and metaphysical naturalism), we can see that what remains in the public education debate is no longer legitimately an issue of church and state, because neither theism per se nor any religious system is involved in the question. Neither design nor randomness can be proven—they are matters of deduction since both are based on a combination of probabilities and metaphysical presuppositions. If randomness cannot be sustained in certain cases, that still does not "prove" design. Likewise, if design cannot be sustained in certain cases, that does not "prove" randomness.[12]

If irreducible complexity is a valid observation, it should not be ignored on the basis of its common association with a design solution. The objective is for public education to inform students of scientifically plausible mechanisms without straying from empirical science into metaphysical teleology or dysteleology, either in what is taught or in what is banned from the classroom.

Various models for descriptive mechanisms of material origins could theoretically be taught, whatever their teleological underpinnings, as long as they have an appropriate level of scientific plausibility as descriptive mechanisms. At present, however, biological evolution is the reigning paradigm. We have proposed that Genesis 1 does not offer a competing descriptive mechanism for material origins, and Intelligent Design likewise does not currently have a replacement model to propose. The Discovery Institute, a think tank that explores Intelligent Design, agrees with this assessment. They do not promote a requirement to teach Intelligent Design.

> Discovery Institute recommends that states and school districts focus on teaching students more about evolutionary theory, including telling them about some of the theory's problems that have been discussed in peer-reviewed science journals. In other words, evolution should be taught as a scientific theory that is open to critical scrutiny, not as a

> sacred dogma that can't be questioned. We believe this is a commonsense approach that will benefit students, teachers, and parents.[13]

On the other hand, the Discovery Institute does not agree with legislation or policy that prohibits teachers from discussing design. "Although Discovery Institute does not advocate requiring the teaching of intelligent design in public schools, it does believe there is nothing unconstitutional about discussing the scientific theory of design in the classroom. In addition, the Institute opposes efforts to persecute individual teachers who may wish to discuss the scientific debate over design."[14] Here it would have to be clarified just what is meant by the "scientific theory of design" beyond being a reference to irreducible complexity. Consequently it should be noted that the Discovery Institute would not agree that teleological models do not belong in the science classroom.

For those concerned with the purity of science, the focus on descriptive mechanisms in an empirical discipline will be welcomed, and considering legitimate weaknesses in the reigning paradigm should pose no problem since science always accepts critiques—that is how it develops and improves. For those concerned about the Bible and the integrity of their theology, the descriptive mechanisms that compose the evolutionary model need not be any more problematic for theology than the descriptive disciplines of meteorology or embryology. These descriptive mechanisms can operate within either a teleological or dysteleological system.[15] If all parties were willing to agree to similar teleological neutrality in the classrooms dedicated to instruction in empirical science, the present conflict could move more easily toward resolution.[16]

In conclusion, when origins are discussed in the classroom, empirical science should be taught. We have discussed three important criteria regarding what constitutes empirical science:

1. It is based on a material ontology and premised on methodological naturalism (this eliminates Genesis from the classroom).
2. It is focused on scientifically valid descriptive mechanisms with their strengths and weaknesses acknowledged. So it should include critiques of Neo-Darwinism as well as other origins theories that are trying to offer better explanations of current observations.
3. It must be teleologically neutral (this rules out Genesis, metaphysical naturalism and design).

Summary of Conclusions

1. Genesis operates primarily within a functional ontology as a faith system.
2. Genesis is insistent in affirming teleology with no possible neutrality.
3. Consequently Genesis should not be taught in empirical science classrooms, for it is not empirical science.
4. Empirical science operates within a material ontology and can be taught as a byproduct of that ontology.
5. Empirical science need not favor teleology or dysteleology and should remain neutral on the issue as much as possible.
6. What science has to offer concerning descriptive mechanisms of material origins can be explored in metaphysically neutral ways without offense to biblical affirmations in Genesis 1.
7. If metaphysical naturalism were to be allowed in the science classroom, then there would no longer be any logical reason to ban discussion of design. Since metaphysical naturalism opposes teleological conclusions, it functions on the same metaphysical plane as design, which opposes dysteleological conclusions.
8. Irreducible complexity has a potential role in the empirical sci-

ence classroom but should not be a matter for legislation one way or the other.

Having granted the role of empiricism in the science classroom, our public educational systems are woefully inadequate if curricula totally ignore metaphysics. I would not want to burden scientists with the task of teaching metaphysics in their science classrooms—whether their metaphysics agree with mine or not. Likewise we need not introduce theology into the public curriculum, though it may have a defensible place as an academic discipline. But somewhere students should be taught about metaphysical systems and the alternatives, and about how a variety of metaphysical systems could integrate with science. This is not an issue of faith, or of a particular religion, or of biblical teaching. It is simply an issue of a well-rounded education. "The only way around this logjam is to *decouple the philosophical (or religious) commitments from the science.*"[17]

The fact is that even though empirical science can be taught as such, scientists must function in an integrated world. A scientist could be at the top of his or her scientific discipline, but that would not mean the scientist was equipped to apply his or her scientific expertise to the various social issues that arise in our world. Bioethics requires an understanding of biology *and* of ethics. Decisions about applied technologies, genetic research, fossil fuel use, environmental controls and a myriad of other important issues require not only scientific training but metaphysical (philosophical and even theological) sophistication. If scientists are the ones making decisions for how their science will find its use in society, they must be as astute in thinking about the metaphysical aspects as they are in thinking about the scientific issues.

It is important that we teach empirical science and teach it well. But empirical science is not an education unto itself that can serve all the needs of society or that can serve as the sum of one's education. The physical sciences are only one branch of education, and we dare not isolate them from the humanities or elevate them as self-contained. As a consequence of these conclusions, I would propose the following resolutions:

Be it resolved:

1. that teachers of science education in the public arena should maintain teleological neutrality to the fullest of their ability;
2. that publishers of science curricula and textbooks for public education should maintain teleological neutrality, and that administrators and science departments should make such neutrality one of the criteria in the selection of textbooks;
3. that administrators in public education should develop courses in which metaphysical options can be considered and that are taught by those who are educated in metaphysics, because it is important for students not only to be competent scientists, but also educated philosophers equipped to make the complex decisions that challenge public policymaking;
4. that people of faith should cease trying to impose their own teleological mandates on public science education; and people who are skeptical of faith should cease trying to impose their own dysteleological mandates on public science education;
5. that those who honor the Bible should allow it to find its theological affirmations as a functional cosmology rather than pressing it into service in public education as if it offered a descriptive mechanism for material origins.

Technical Support

Fowler, Thomas B., and Daniel Kuebler. *The Evolution Controversy: A Survey of Competing Theories*. Grand Rapids: Baker Academic, 2007.

Summary and Conclusions

The purpose of this book has been to introduce the reader to a careful reconsideration of the nature of Genesis 1. I have proposed that the most careful, responsible reading of the text will proceed with the understanding that it is ancient literature, not modern science. When we read the text in the context of the ancient world we discover that what the author truly intended to communicate, and what his audience would have clearly understood, is far different from what has been traditionally understood about the passage.

The position that I have proposed regarding Genesis 1 may be designated the *cosmic temple inauguration* view. This label picks up the most important aspect of the view: that the cosmos is being given its functions as God's temple, where he has taken up his residence and from where he runs the cosmos. This world is his headquarters.

The most distinguishing feature of this view is the suggestion that, as in the rest of the ancient world, the Israelites were much more attuned to the functions of the cosmos than to the material of the cosmos. The functions of the world were more important to them and more interesting to them. They had little concern for the material structures; significance lay in who was in charge and made it work. As a result, Genesis 1 has been presented as an account of functional origins (specifically functioning for people) rather than an account of material origins (as we have been generally inclined to read it). As an account of functional origins, it offers no clear information about material origins.

The key features of this interpretation include most prominently:

- The Hebrew word translated "create" *(bārāʾ)* concerns assigning functions.
- The account begins in verse 2 with no functions (rather than with no material).
- The first three days pertain to the three major functions of life: time, weather, food.
- Days four to six pertain to functionaries in the cosmos being assigned their roles and spheres.
- The recurring comment that "it is good" refers to functionality (relative to people).
- The temple aspect is evident in the climax of day seven when God rests—an activity in a temple.

The account can then be seen to be a seven-day inauguration of the cosmic temple, setting up its functions for the benefit of humanity, with God dwelling in relationship with his creatures.

This proposed reading of Genesis 1 then led to a consideration of the implications for thinking about theology, evolution and Intelligent Design. If Genesis 1 is not an account of material origins, then it offers no mechanism for material origins, and we may safely look to science to consider what it suggests for such mechanisms. We may find the theories proposed by scientists to be convincing or not, but we cannot on the basis of Genesis 1 object to any mechanism they offer. The theological key is that whatever science proposes that is deemed substantial, our response is, "Fine, that helps me see the handiwork of God." Accepting at least some of the components of biological evolution as representing the handiwork of God, we could propose a mechanism for material origins designated *teleological evolution* meaning that evolutionary processes may well describe some aspects of origins (noting that human origins need to be discussed separately), even though much controversy still exists about how evolutionary changes took place. The use of the adjective *teleological* differentiates this view from standard Neo-Darwinism, as teleology affirms the conviction that the process understands material origins as God's creative work with a purpose and a goal. Consequently we are not surprised that there are evidences of design.

We proposed that this view is not only exegetically sound, it is also theologically robust and actually strengthens our theology of creation. With confidence in reading Genesis 1 as supported by the original context, and the confidence in the theological vibrancy of our commitment, we have discovered several advantages:

1. When discussing our faith with skeptics, we need not fear the science discussion. We can relax and respond to any proposal they make with, "Yes, but there is no reason God could not have been involved in that process." The supposed conflict between science and faith is often simply a misunderstanding. There is, in fact, evidence that the conflict was promoted from the science side before it was ever taken up from the faith side.[1]

2. A second advantage is that by holding the cosmic temple inauguration view of Genesis and the teleological evolution view of material origins we may be able to curb the constant

attrition of faith that takes place as students interested in science have been told that they have to choose between science and faith. Such a choice is not necessary.

3. A third advantage is that we may begin refocusing our concerns about public education. Rather than trying to push the agenda that young-earth creationism or Intelligent Design needs to be taught in the schools, we can focus on demanding that metaphysical naturalism, a matter of belief rather than science, *not* be bundled together with the teaching of evolution. We can call schools, teachers and textbook publishers to account for the ways that they insert dysteleology (which is not science, but belief) into the curriculum. Furthermore public education should be interested in teaching evolution with all of its warts and problems, and not overstating the case.

The concern of this book is neither to tell scientists how they should or should not do science, nor to determine what scientific conclusions are right or wrong. It should be noted that this book is *not* promoting evolution. The issue I have attempted to approach concerns what scientific ideas or conclusions that the believer who wants to take the Genesis account seriously is obliged to reject. Is there *science* that is unacceptable in biblical/theological terms? Or are only the metaphysical implications adopted by some scientists unacceptable? Is it the Genesis account that serious scientists are compelled to reject? Or only the implications of some traditional interpretations? Biological evolution is the reigning paradigm, so we have asked whether this view requires the believer to compromise theology or biblical teaching. We have concluded that there is nothing intrinsic to the scientific details (differentiated from the metaphysical implications that some draw) that would require compromise.

Scientists should be committed to refining, modifying and

even overhauling or overthrowing any reigning paradigm that is proven inadequate. This is the nature of scientific inquiry. Having said that, whatever aspects of evolution that continue to provide the best explanation for what we observe should not, in most cases, be objectionable for Christians. In promoting the theological position in the Bible and the interpretation of Genesis 1 presented here, there is no reason to believe that biological evolution teaches something contradictory to the Bible (though some evolutionists are proponents of metaphysical conclusions that contradict the Bible). Believing in the Bible does not require us to reject the findings of biological evolution, though neither does it give us reason to promote biological evolution. Biological evolution is not the enemy of the Bible and theology; it is superfluous to the Bible and theology. The same could be said for the big bang and for the fossil record.

The view presented here presents a way forward through the morass created by the entrenched positions of Neo-Darwinian evolution and the commitment to Scripture and sound theology. The problem is well articulated by Fowler and Kuebler:

> The ante has been raised so high by the polemical nature of the controversy that resolution in favor of one school will have catastrophic implications for the other. On the one hand, the scientific community by and large, including the National Academy of Sciences, has staked the prestige of science on a particular theory with considerable explanatory power but known problems, in part because it is consistent with a naturalistic philosophy. On the other hand, Creationists have for all intents and purposes staked the truth of their religion on the falsity of that same theory, because of the perceived need for a literal interpretation of the Bible. Clearly, neither the proponents of Creationism nor those of

> Neo-Darwinism can permit their side to lose or even give ground, regardless of the facts; the extra-scientific stakes for both are just too high.[2]

In the view presented in this book, neither camp must "give ground," but they both need to be willing to let go of their polemical antagonism. Neo-Darwinism proponents need not make any concessions about what empirical science proposes for material origins. They only have to stop promoting dysteleology as if it were an essential corollary to the science. They also have to stop acting as if Neo-Darwinism has no flaws and no need of modification.[3] Creationists need not give up their theology of God's total involvement in creation, nor do they need give up a "literal" reading of Genesis 1. They only have to acknowledge that traditional interpretations or understandings of English words do not necessarily constitute the most faithful reading of the text. We are *not* proposing that readers of the Bible back off to a figurative or simply literary reading of Genesis 1. We would suggest, instead, that the reading this book proposes is precisely what the Genesis author and audience would have understood.

Finally, both sides need to give up their stubborn antagonism. As Gerald Runkle writes in his book *Good Thinking:*

> It is the mark of stubborn and dogmatic persons to be oblivious to the need either to test their own beliefs or to recognize the successful tests that opposing beliefs have undergone. Copernicus caused widespread consternation when he suggested that the earth revolved around the sun. Though he had impressive evidence for his theory, it was received in ill humor by most religious groups. Martin Luther complained: "People give ear to an upstart astrologer who strove to show that the earth revolved, not the heavens or the firmament, the sun and the moon. . . . This fool wishes to re-

verse the entire science of astronomy; but the sacred Scripture tells us that Joshua commanded the sun to stand still and not the earth."[4]

We must keep in mind that we are presumptuous if we consider our interpretations of Scripture to have the same authority as Scripture itself. Nobody is an infallible interpreter, and we must always stand ready to reconsider our interpretations in light of new information. We must not let our interpretations stand in the place of Scripture's authority and thus risk misrepresenting God's revelation. We are willing to bind reason if our faith calls for belief where reason fails. But we are also people who in faith seek learning. What we learn may cause us to reconsider interpretations of Scripture, but need never cause us to question the intrinsic authority or nature of Scripture.

FAQs

Q: When and how did God create the material world?
A: According to the interpretation offered in this book, the Bible does not tell us, so we are left to figure it out as best we can with the intellectual capacity and other tools that God gave us. But the material world was created by him.

Q: Where do the dinosaurs and fossil "homo" specimens fit in?
A: In the view presented in this book, these creatures could be part of the prefunctional cosmos—part of the long stage of development that I would include in the material phase. Since the material phase precedes the seven days of Genesis 1, these would all be relegated to the obscure and distant past. The anthropological specimens would not be viewed as humans in the image of God. They would not be assessed morally (any more than an animal would), and they were subject to death as any animal was. Most did not survive alongside the humans that the Bible discusses, and others would have died off early.

Q: Isn't this just really a dodge to accommodate evolution?

A: The interpretation set forth in this book arose out of my desire to fully understand the biblical text. Understanding evolution and its role is a much lower value. Evolution represents the current scientific consensus to explain the many observations that have been made in paleontology, genetics, zoology, biochemistry, ecology and so on. The question is how much of what is involved in biological evolution runs counter to what I understand to be biblical claims and theological realities. In the interpretation of the text that I have offered, very little found in evolutionary theory would be objectionable, though certainly some of the metaphysical claims of evolution remain unacceptable.

Q: Why don't you want to just read the text literally?
A: I believe that this *is* a literal reading. A literal reading requires an understanding of the Hebrew language and the Israelite culture. I believe that the reading that I have offered is the most literal reading possible at this point. Someone who claims a "literal" reading based on their thinking about the English word "create" may not be reading the text literally at all, because the English word is of little significance in the discussion.

Q: What would people have seen if they were there as eyewitnesses (i.e., what "really happened") on these days?
A: We overrate eyewitnesses in our culture. The Bible is much more interested in understanding what God did rather than what an eyewitness would see. For example, an eyewitness would have seen the waters of the Red Sea part, but would have no physical evidence that God did it. Genesis 1 is an account of creation intended to convey realities about the origins of the cosmos and God's role in it and his purpose for it. Most importantly it is designed to help the reader understand that the cosmos should be understood as a temple that God has set up to operate for people

as he dwells in their midst. The perspective of an eyewitness would be inadequate and too limited to be of any good. Genesis 1 is not intended to be an eyewitness account.

Q: Why can't Genesis 1 be both functional and material?
A: Theoretically it could be both. But assuming that we simply *must* have a material account if we are going to say anything meaningful is cultural imperialism. We cannot demand that the text speak to us in our terms. Just as we cannot demand a material account, we cannot assume a material account just because that is most natural to us and answers the questions we most desire to ask. We must look to the text to inform us of its perspective. In my judgment, there is little in the text that commends it as a material account and much that speaks against it. (See pp. 93-94.)

Q: If this is the "right" reading, why didn't we know about it until now?
A: While this reading is initially based on observations from the biblical text (as opposed to observations about the ancient worldview), without an understanding of the ancient worldview, it would have been difficult to ask the questions that have led to this position and nearly impossible to provide the answers to the questions that we have proposed. The worldview of antiquity was lost to us as thinking changed over thousands of years, and the language and literature of the ancient world was buried in the sands of the Middle East. It was only with the decipherment of the ancient languages and the recovery of their texts that windows were again opened to an understanding of an ancient worldview that was the backdrop of the biblical world. This literature and the resulting knowledge has made it possible to recover the ways of thinking that were prominent in the ancient world and has given us new insight into some difficult biblical texts (see my *Ancient*

Near Eastern Thought and the Old Testament [Grand Rapids: Baker Academic, 2006]).

Q: Why would God make it so difficult for me to understand his Word?
A: Given God's decision to communicate, he had to choose one language and culture to communicate to, which means that every other language and culture has their work cut out for them. As readers from a different language and culture, we have to try to penetrate the original language and culture if we are to receive the maximum benefits of God's revelation. We also need to seek greater understanding when we are confronted with information from outside the Bible (whether ancient or modern) and want to figure out how it integrates into what we believe the Bible is saying. It is relieving to recognize that the basics of God's revelation of himself (including his Creator role) are easily skimmed off the surface, but it is not surprising that God's Word contains infinite depth and that it should require constant attention to study with all the tools we have available. God is not superficial, and we should expect that knowledge of him and his Word would be mined rather than simply absorbed. This means that all of us will be dependent on others with particular skills to help us succeed in the enterprise of interpretation. This is not elitism; it is the interdependence of the people of God as they work together in community to serve one another with the gifts they have.

Q: How can this view of Genesis be taught to children in Sunday school and Christian elementary schools?
A: The most important aspects of Genesis 1 to emphasize for children is that God was involved at every level and that he is responsible for setting up the world so that it works. This is the theological side of the question. On the textual side of the question, when

Genesis 1 is the basis for a Bible story, we can emphasize what is most important: functions and operations. The teacher would not need to get into the issue of Genesis 1 not being an account of material origins. That could come at later levels of study. It would be important, however, not to criticize evolution as contradictory to the Bible. Rather statements can be made that whatever processes were involved, God was controlling those processes.

Notes

Proposition 1: Genesis 1 Is Ancient Cosmology

[1]For examples of ancient thought in numerous categories of science, see Denis Lamoureux, *Evolutionary Creation* (Eugene, Ore.: Wipf and Stock, 2008), pp. 105-47.

[2]One of the most common examples given by those who suggest there is a latent scientific consideration is that Is 40:22 posits a spherical earth. This cannot be sustained because its terminology only indicates a disk, not a sphere.

[3]Richard Bube, *The Human Quest* (Waco, Tex.: Word, 1971), pp. 26-27.

[4]See the contrast between the extremes of deism and micromanagement discussed in Terence E. Fretheim, *God and World in the Old Testament: A Relational Theology of Creation* (Nashville: Abingdon, 2005), pp. 7, 22-24.

[5]This observation came from my student Jeremey Houlton.

Proposition 2: Ancient Cosmology Is Function Oriented

[1]For more extensive summary and discussion, see John Walton, "Creation," in *Dictionary of the Old Testament: Pentateuch,* ed. T. Desmond Alexander and David W. Baker (Downers Grove, Ill.: InterVarsity Press, 2003), pp. 155-68.

[2]For a good treatment of the ancient Near Eastern creation texts see Richard Clifford, *Creation Accounts in the Ancient Near East and in the Bible,* Catholic Biblical Quarterly Monograph Series 26 (Washington, D.C.: Catholic Biblical Association, 1994).

[3]Ibid., p. 28, translated into English from J. van Dijk's French translation in "Existe-t-il un 'Poème de la Création' Sumérien?" in *Kramer Anniversary Volume: Cuneiform Studies in Honor of Samuel Noah Kramer,* ed. B. Eichler et al. (Neukirchen-Vluyn: Butzon & Bercker, 1976), pp. 125-33.

[4]James P. Allen, *Genesis in Egypt: The Philosophy of Ancient Egyptian Creation Accounts* (New Haven, Conn.: Yale Egyptological Seminar, 1988), pp. 57-58: "Creation is the process through which the One became the Many."

[5]Coffin Texts, spell 76, translation by James Allen, in *Context of Scripture* 1.6, ed. W. Hallo and K. L. Younger (Leiden: Brill, 1997), p. 10.

[6]Ibid., p. 16.

[7]Harry A. Hoffner Jr., "Song of Ullikummi," in *Hittite Myths,* Society of Biblical Literature Writings from the Ancient World 2 (Atlanta: Scholars Press, 1990), p. 59, §61. The speaker is Ubelluri, a god similar to Atlas in Greek mythology, who holds up the cosmos from his place in the netherworld.

[8]Translation from Miriam Lichtheim, *Ancient Egyptian Literature* (Berkeley: University of California Press, 1980), 3:210-11.

[9]See Benjamin R. Foster, *Before the Muses: An Anthology of Akkadian Literature,* 3rd ed. (Bethesda, Md.: CDL Press, 2005), p. 464; Wayne Horowitz, *Mesopotamian Cosmic Geography* (Winona Lake, Ind.: Eisenbrauns, 1998), pp. 117-18.

[10]For this interpretation see Horowitz, *Mesopotamian Cosmic Geography,* p. 117.

[11]*The Debate Between Winter and Summer* 5.3.3, lines 1-11 <etcsl.orinst.ox.ac.uk>.

[12]For the machine vs. kingdom contrast see John Stek, "What Says the Scripture?" in *Portraits of Creation,* ed. H. J. van Till (Grand Rapids: Eerdmans, 1990), p. 255.

Proposition 3: "Create" (Hebrew *bārāʾ*) Concerns Functions

[1]One might claim that this puts us at the mercy of Hebrew scholars, but remember, it was Hebrew scholars who gave us the English verb "create" to begin with in our translations, so nothing has changed, we have just faced reality.

[2]From a practical standpoint, we know that this is true. Unfortunately, sometimes when we get to scholarly analysis we forget how the world of words generally works and try to use etymology rather than usage, even though we know that in the language we speak, etymology is an unreliable guide to meaning. We know that "awful" does not mean "full of awe" and that "understand" does not mean "to stand under." We must resist the temptation to use etymology in word analysis. The only reliable guide is usage.

[3]For a discussion with examples and a bit more linguistic detail see John Walton, "Principles for Productive Word Study," in *The New International Dictionary of Old Testament Theology and Exegesis,* ed. W. VanGemeren (Grand Rapids: Zondervan, 1997), 1:161-71.

[4]Direct objects in the Dead Sea Scrolls include: vault, light, morning, evening, age, spirit, spice, treasury, sanctuary, people, deed, righteous one, wicked one, flesh, evil and shame. See full citations in *Dictionary of Classical Hebrew,* ed. D. J. A. Clines (Sheffield, U.K.: Sheffield Academic Press, 1993-2001), 2:258-59; and discussion in H. Ringgren, "ברא *Baraʾ*," *Theological Dictionary of the Old Testament* (Grand Rapids: Eerdmans, 1974-), 2:249. The study of the objects with similar conclusions was done by John Stek, "What Says the Scripture?" in *Portraits of Creation,* ed. H. J. van Till (Grand Rapids: Eerdmans, 1990), pp. 203-65, see particularly p. 208. The conclusion he reaches is that "In biblical language, *baraʾ* affirms of some existent reality *only that God conceived, willed, and effected it*" (p. 213). He also catalogs biblical references where *bārāʾ* involves providential processes over time (p. 212).

[5]See Stek, "What Says the Scripture?" pp. 203-65 (especially p. 208).

[6]It should be noted, however, that in a large percentage of the cases where the usage is ambiguous, a further explanation is offered that

indicates a functional interest (noted in the last column).

[7]Our discussion here can only be summary. For detailed discussion see John Walton, *Genesis,* NIV Application Commentary (Grand Rapids: Zondervan, 2001), pp. 67-70; John Walton, *Genesis One as Ancient Cosmology* (Winona Lake, Ind.: Eisenbrauns, forthcoming).

[8]John Sailhamer, *Genesis Unbound* (Sisters, Ore.: Multnomah, 1996), p. 38. Detailed discussion may be found in Sailhamer's Genesis commentary in the *Expositor's Bible Commentary,* ed. F. E. Gaebelein (Grand Rapids: Zondervan, 1990), 2:20-23, and a summary by Bill Arnold in the article on *rēšît* in *New International Dictionary of Old Testament Theology and Exegesis,* ed. W. A. VanGemeren (Grand Rapids: Zondervan, 1997), 3:1025-26.

Proposition 4: The Beginning State in Genesis 1 Is Nonfunctional

[1]David Tsumura, *Creation and Destruction* (Winona Lake, Ind.: Eisenbrauns, 2005), p. 35.

[2]The word that NIV renders "northern (skies)" is *ṣāpôn,* the Hebrew word for "north" by virtue of Mt. Zaphon, which is in the north (see Ps 48:2 and Is 14:13). More importantly, it refers to the place where the divine council meets and therefore serves as a reference to heaven. This is confirmed by the use of the verb "stretched out," which in cosmological texts in the Bible is an activity connected to the heavens. So as the North (the place where the heavenly assembly meets) is stretched out over *tōhû,* the earth is suspended over X (NIV "nothing"). Psalm 104:2-3 indicates that the heavens are stretched out over the heavenly cosmic waters (the waters above). Psalm 24:1-2 tells us that the earth is founded on the cosmic waters (cf. Ps 136:6).

[3]Erik Hornung, *Conceptions of God in Ancient Egypt* (Ithaca, N.Y.: Cornell University Press, 1982), pp. 174-76.

[4]Ibid., p. 177.

[5]Ibid., p. 171; Siegfried Morenz, *Egyptian Religion* (Ithaca, N.Y.: Cornell University Press, 1973), p. 173. Texts include Pyramid Text 1208c (Morenz, *Egyptian Religion,* p. 173); Coffin Texts 4, 36 (spell 286) (Morenz, *Egyptian Religion,* p. 173); Heliopolis (Morenz, *Egyptian*

Religion, p. 173); Stele Leyden 5.12 (Morenz, *Egyptian Religion,* p. 173); "Ptah, Lord of maat . . . who lifted up the sky and created things that be" (Morenz, *Egyptian Religion,* p. 174); Memphite Theology, line 14: Ptah, creating through the Ennead, is identified as the one who "pronounced the identity of everything."

[6]NBC refers to the Nies Babylonian Collection, from Yale University.

[7]Richard Clifford, *Creation Accounts in the Ancient Near East and in the Bible,* Catholic Biblical Quarterly Monograph Series 26 (Washington, D.C.: Catholic Biblical Association, 1994), p. 28; translated into English from J. van Dijk's French translation in "Existe-t-il un 'Poème de la Création' Sumérien?" in *Kramer Anniversary Volume: Cuneiform Studies in Honor of Samuel Noah Kramer,* ed. B. Eichler et al. (Neukirchen-Vluyn: Butzon & Bercker, 1976), pp. 125-33.

Proposition 5: Days One to Three in Genesis 1 Establish Functions

[1]See extensive discussion of all of the different categories of metonymy and the biblical occurrences in E. W. Bullinger, *Figures of Speech Used in the Bible* (Grand Rapids: Baker, 1968), pp. 538-608.

[2]This makes even more sense when we recognize that darkness is not an object either to us or in the ancient world.

[3]P. Seely, "The Firmament and the Water Above," *Westminster Theological Journal* 54 (1992): 31-46.

[4]Papyrus Insinger, *Ancient Egyptian Literature,* trans. Miriam Lichtheim (Berkeley: University of California Press, 1980), 3:210.

[5]See Benjamin R. Foster, *Before the Muses: An Anthology of Akkadian Literature,* 3rd ed. (Bethesda, Md.: CDL Press, 2005), pp. 436-86.

[6]John Walton, *Genesis,* NIV Application Commentary (Grand Rapids: Zondervan, 2001), pp. 344-45.

Proposition 6: Days Four to Six in Genesis 1 Install Functionaries

[1]For further discussion see John Walton, *Genesis,* NIV Application Commentary (Grand Rapids: Zondervan, 2001), pp. 122-23, drawing

on the work of W. Vogels, "The Cultic and Civil Calendars of the Fourth Day of Creation (Gen 1,14b)," *Scandinavian Journal of the Old Testament* 11 (1997): 163-80.

[2]Richard Clifford, *Creation Accounts in the Ancient Near East and in the Bible,* Catholic Biblical Quarterly Monograph Series 26 (Washington, D.C.: Catholic Biblical Association, 1994), p. 67. In the Akkadian version the three named gods charge the great astral gods to produce day and to assure the regular sequence of months for astrological observation.

[3]I am grateful to my student Liesel Mindrebo for pointing out this pattern. Important other uses of this verb in cosmology contexts can be found in Ex 34:10; 1 Kings 12:32-33; Job 9:9; Is 41:17-20; 45:7; Jer 38:16; see Walton, *Genesis,* pp. 124-25. For my detailed lexical analysis of this verb, see John Walton, *Genesis One as Ancient Cosmology* (Winona Lake, Ind.: Eisenbrauns, forthcoming).

[4]The Exploits of Ninurta 1.6.2 <etcsl.orinst.ox.ac.uk>.

[5]It should be noted that the function of an archetype does not rule out their historical (or biological) reality. In Romans 5 Jesus stands as an archetype alongside Adam. Abraham is identified as an archetype of people of faith. These are historical figures who are being used in the literature for their archetypal significance.

Proposition 7: Divine Rest Is in a Temple

[1]In Ex 31:17 there is also an indication that God "refreshed himself."

[2]Temple Hymn of Keš 4.80.2, D.58A-F <etcsl.orinst.ox.ac.uk>.

[3]Translations from *The Context of Scripture,* ed. W. Hallo and K. L. Younger (Leiden: Brill, 1997), 1:111.

[4]Notice also how all the set-up tasks referred to are functional rather than material in nature. That is, there was no discussion of the material phase of manufacturing the tower or the cables, or of writing the software.

Proposition 8: The Cosmos Is a Temple

[1]It is difficult to date the piece. The copy is Seleucid (Richard Clifford, *Creation Accounts in the Ancient Near East and in the Bible,* Catholic

Biblical Quarterly Monograph Series 26 [Washington, D.C.: Catholic Biblical Association, 1994], p. 62), but Wayne Horowitz considers it to derive from a Sumerian original (*Mesopotamian Cosmic Geography* [Winona Lake, Ind.: Eisenbrauns, 1998], pp. 129-31).

[2]Benjamin R. Foster, *Before the Muses: An Anthology of Akkadian Literature,* 3rd ed. (Bethesda, Md.: CDL Press, 2005), p. 488.

[3]Richard Clifford, *Creation Accounts in the Ancient Near East and in the Bible,* Catholic Biblical Quarterly Monograph Series 26 (Washington, D.C.: Catholic Biblical Association, 1994), p. 61.

[4]J. Black et al., *The Literature of Ancient Sumer* (Oxford: Oxford University Press, 2004), pp. 325-30, lines 13-16 cited, but the ideas are repeated throughout the piece.

[5]Gudea B.xx.8-11 translated by Thorkild Jacobsen, *The Harps That Once . . .* (New Haven, Conn.: Yale University Press, 1987), pp. 441-42.

[6]Jan Assmann, *The Search for God in Ancient Egypt* (Ithaca, N.Y.: Cornell University Press, 2001), p. 38; Cf. Clifford, *Creation Accounts,* pp. 105-6; See also John Lundquist, "What Is a Temple? A Preliminary Typology," in *The Quest for the Kingdom of God,* ed. H. Huffman (Winona Lake, Ind.: Eisenbrauns, 1983), p. 208; Othmar Keel, *Symbolism of the Biblical World* (New York: Seabury Press, 1978), p. 113, indicates that this is true of both Egypt and Mesopotamia.

[7]Temple Hymns 4.80.1 <etcsl.orinst.ox.ac.uk>.

[8]Assmann, *Search for God,* p. 37 (italics in original).

[9]Ibid., pp. 35-36.

[10]L. R. Fisher, "Creation at Ugarit and in the Old Testament," *Vetus Testamentum* 15 (1965): 320.

[11]Josephus *The Jewish War* 3, 7.7, trans. H. St. J. Thackery, Loeb Classical Library (Cambridge, Mass.: Harvard University Press, 1957), p. 403.

[12]Jon Levenson, "The Temple and the World," *Journal of Religion* 64 (1984): 295, even suggests tentatively that there is some possibility that the temple in Jerusalem may have been called by the name "Heaven and Earth." Temples in the ancient world had names, and many of them refer to the temple's cosmic significance, e.g., the Temple at Nippur, "Duranki," which means, "Bond of Heaven and Earth," and at

Babylon, "Etemenanki," which means, "Foundation of Heaven and Earth." This could be supported from verses such as Is 65:17-18 in which a creation of a new heaven and new earth is paralleled by creating Jerusalem, soon followed up in Is 66:1 with the picture of the cosmos as God's temple. The idea was suggested to Levenson in an article by G. Ahlstrom, "'Heaven on Earth'—at Hazor and Arad," in *Religious Syncretism in Antiquity,* ed. B. A. Pearson (Missoula, Mont.: Scholars Press, 1975), pp. 67-83. If this view could be substantiated, Gen 1:1 would take on a new level of meaning as a reference to the cosmic temple.

[13]Gordon J. Wenham, "Sanctuary Symbolism in the Garden of Eden Story," in *Proceedings of the Ninth World Congress of Jewish Studies* (Jerusalem: World Union of Jewish Studies, 1986), p. 19.

[14]Examples include the façade of the temple of Inanna in Uruk, which pictures guardian beings surrounded by the flow of streams; the investiture fresco at Mari, and many statues that show the individual holding a jar from which waters flow.

[15]Victor Hurowitz, *I Have Built You an Exalted House,* Journal for the Study of the Old Testament Supplement Series 115 (Sheffield, U.K.: JSOT Press, 1992), p. 242.

[16]Levenson, "Temple and the World," points out these examples, and in Isaiah 6 he goes even further to suggest that the word here translated "full" is not an adjective, but a noun, "fullness," in which case the proper translation should be "The fullness of the whole earth is his glory" (p. 289).

Proposition 9: The Sevens Days of Genesis 1 Relate to the Cosmic Temple Inauguration

[1]Jon Levenson, "The Temple and the World," *Journal of Religion* 64 (1984): 288-89; Victor Hurowitz, *I Have Built You an Exalted House,* Journal for the Study of the Old Testament Supplement Series 115 (Sheffield, U.K.: JSOT Press, 1992), pp. 260-61, 275-76. The number seven is prevalent, though variations appear (e.g., Esarhaddon's dedication of his temple in Assur over three days, and Assurnasirpal's

dedication of Kalhu over ten days). Hurowitz's appendix on pp. 280-82 provides the entire list of over forty texts. Another striking seven-day festival is an Old Babylonian ritual from Larsa, see See E. C. Kingsbury, "A Seven Day Ritual in the Old Babylonian Cult at Larsa," *Hebrew Union College Annual* 34 (1963): 1-34. There is no evidence that this is a temple dedication ritual; in fact each day focuses on a different god. Intriguingly the rituals for each new day also begin in the evening (p. 26). On p. 27 Kingsbury lists several other seven-day rituals.

[2]Most of Gudea Cylinder B is taken up with the installation of functionaries.

[3]Gudea B.vi.11-16, translation by R. Averbeck, from *The Context of Scripture,* ed. W. Hallo and K. L. Younger (Leiden: Brill, 1997), 2:155.

[4]Moshe Weinfeld, "Sabbath, Temple and the Enthronement of the Lord—The Problem of the Sitz im Leben of Genesis 1.1—2.3," in *Mélanges bibliques et orientaux en l'honneur de M. Henri Cazelles,* ed. A. Caquot and M. Delcor, Alter Orient und Altes Testament 212 (Neukirchener-Vluyn: Neukirchener; Kevelaer: Butzon & Bercker, 1981), pp. 502-12.

[5]Concordism attempts to read modern scientific meaning into the ancient words and texts. We will discuss the hermeneutical problems with this approach, that is, its problems with interpreting the biblical text, see pp. 104-7.

Proposition 10: The Seven Days of Genesis 1 Do Not Concern Material Origins

[1]See pp. 53, 60.

[2]Some might contend that the Hebrew verb *ʿāśâ* ("make" vv. 7, 16, 25, 26) and *nātan* ("set" v. 17) provide evidence for the material nature of the text. These discussions are more complex and will be treated at length in John Walton, *Genesis One as Ancient Cosmology* (Winona Lake, Ind.: Eisenbrauns, forthcoming). To summarize, *ʿāśa* is often translated "do" (e.g., one's business), and the evidence favors that understanding here

(cf. the use in Ex 20:8-11). In similar fashion, *nātan* often means "appoint," and that suits this context well.

[3]Even more questionable would be the decision to oppose the possibility of an old earth simply because that would give time for evolution. That would be folly—evolution would need to stand or fall on its own merits.

[4]This would be similar to the fact that when God said "Do not murder" on Mt. Sinai, it is not as if before then everyone murdered whomever they wanted. It was not that the law was new but that it was put in a new context of God's covenant. In a similar manner, it is not that the sun was not previously shining but now it is seen in a different context—the context of the cosmic temple.

Proposition 12: Other Theories of Genesis 1 Either Go Too Far or Not Far Enough

[1]For a balanced popular treatment see Gordon Glover, *Beyond the Firmament* (Chesapeake, Va.: Watertree Press, 2007); for a more in-depth look at the strengths and weaknesses of the position and the scientific challenges it faces, see Thomas B. Fowler and Daniel Kuebler, *The Evolution Controversy: A Survey of Competing Theories* (Grand Rapids: Baker Academic, 2007).

[2]Hugh Ross, *The Genesis Question: Scientific Advances and the Accuracy of Genesis* (Colorado Springs, Colo.: NavPress, 2001), p. 9.

[3]Ibid., pp. 24-34.

[4]Ibid., p. 65. Against others in this camp, Ross believes that the sequence of the seven days can be sustained in the scientific events of cosmology.

[5]C. John Collins, *Genesis 1—4* (Phillipsburg, N.J.: P & R, 2006), p. 73.

[6]For a fuller presentation of the framework hypothesis as well as a fair analysis of the other positions, see Henri Blocher, *In the Beginning* (Downers Grove, Ill.: InterVarsity Press, 1984), pp. 39-59.

[7]The various possibilities are presented and analyzed by D. Young, "The Antiquity and the Unity of the Human Race Revisited," *Chris-*

tian Scholar's Review 24, no. 4 (1995): 380-96.

Proposition 13: The Difference Between Origin Accounts in Science and Scripture Is Metaphysical in Nature

[1]This modern distinction was especially championed and articulated by the eighteenth-century philosopher Immanuel Kant.

[2]I wish to thank my colleague Lynn Cohick for this suggestion.

[3]One of the places where this analogy breaks down is that it risks suggesting too distinct a divide between the two layers where no such divide truly exists. Instead the two are fully integrated and in some ways might more resemble a marble cake.

[4]These distinctions are discussed in detail in Denis Lamoureux, *Evolutionary Creation* (Eugene, Ore.: Wipf and Stock, 2008), pp. 69-70.

[5]Perhaps some might claim that the Intelligent Design movement attempts precisely that. This will be discussed in another chapter.

[6]Some may feel that "empirical science" is redundant, but I use the combination just to be sure that I am clear. By *empiricism* I am trying to isolate those aspects of science which value an evidentiary base and seek to focus on that base. In that sense it is distinct from rationalism, though empirical science has always left room for and indeed encouraged rational deductions that are made from an evidentiary base. So, for instance, observations concerning a given artifact may indeed lead to the logical deduction that it was made with a purpose. In a sense this could be an empirical deduction.

[7]For additional discussion and a distinction between "teleological evolution" and "dysteleological evolution" see Lamoureux, *Evolutionary Creation*, pp. 4-5.

[8]Materialism is the view that the material is all there is (bottom layer only). Naturalism describes a cause-and-effect process in scientific terms, with the natural laws as the foundation. Naturalism describes the operation of the bottom layer (sometimes referred to as *methodological naturalism*). Materialism says the bottom layer is all there is (sometimes referred to as *metaphysical naturalism*). Christians need not deny naturalistic operations, but they denounce materialism.

[9]Though the text offers a view of God initially establishing functions in the past, even in that regard its focus is the present and the ongoing future.

Proposition 14: God's Roles as Creator and Sustainer Are Less Different Than We Have Thought

[1]I am grateful to my colleague Robert Bishop for these observations.

[2]Not unlike the ancient Egyptian view in which it happened again each day, though even they differentiated the events on what they referred to as the "first occasion."

[3]Terence Fretheim speaks of a beginning (Originating Creation), a middle (Continuing Creation) and an end (Completing Creation) (Terence E. Fret-heim, *God and World in the Old Testament: A Relational Theology of Creation* [Nashville: Abingdon, 2005], pp. 5-9).

[4]Jürgen Moltmann, *God in Creation: A New Theology of Creation and the Spirit of God* (San Francisco: Harper & Row, 1985), summarized and critiqued by Francis Watson, *Text and Truth* (Grand Rapids: Eerdmans, 1997), pp. 227-36.

[5]Observations and questions posed by Watson, *Text and Truth,* pp. 226-27. In Watson's belief that the "beginning" must be an absolute beginning, he does not consider the possibility that the absolute beginning should be viewed against a functional ontology instead of against a material ontology. This could make a big difference to the implications of the assertion.

[6]This is not an attempt to promote "natural theology," which explores whether God can be perceived in nature without the aid of special revelation. We are unconcerned here with the revelation question as we affirm only that God is at work sustaining the world, however that may be perceived through observation.

[7]Fretheim, *God and World in the Old Testament,* p. 5.

[8]Watson, *Text and Truth,* p. 228.

[9]John Stek, "What Says the Scripture?" in *Portraits of Creation,* ed. H. J. van Till (Grand Rapids: Eerdmans, 1990), pp. 203-65, quote on p. 211. On pp. 242-50 Stek looks in detail at the theological (Reformed)

traditions that have insisted on a sharp break between creation and providence. He points out that their theological concerns are clear as they have sought to insulate God from being the author of evil. But he then points out many biblical texts that show that the Old Testament is more inclined to merge the two (p. 246).

[10]The distinction between *evolution* and *evolutionism* goes as least as far back as C. S. Lewis, "The Funeral of a Great Myth," in *Christian Reflections* (Grand Rapids: Eerdmans, 1967), pp. 82-93 (see especially p. 83). Thus we might suggest that it is not creation and evolution that are at odds, but their ideological cousins, Creationism and Evolutionism.

Proposition 15: Current Debate About Intelligent Design Ultimately Concerns Purpose

[1]Orson Scott Card, *The Call of Earth* (New York: Tor, 1993), p. 138.

[2]There is also a significant mathematical element to their position; see William Dembski, *Intelligent Design* (Downers Grove, Ill.: InterVarsity Press, 1999).

[3]There are alternatives out there such as S. Kauffman, *At Home in the Universe: The Search for the Laws of Self-Organization and Complexity* (Oxford: Oxford University Press, 1996). Kauffman proposes that matter self-organizes, thus making design an expected result intrinsic to the nature of matter and not dependent on a designer.

[4]Thomas B. Fowler and Daniel Kuebler, *The Evolution Controversy* (Grand Rapids: Baker Academic, 2007), pp. 240, 271.

[5]Ibid., p. 237.

[6]Ibid., p. 244.

[7]Ibid., pp. 277-326.

Proposition 16: Scientific Explanations of Origins Can Be Viewed in Light of Purpose, and If So Are Unobjectionable

[1]These have been referred to as "suboptimal." See the discussion in Denis Lamoureux, *Evolutionary Creation* (Eugene, Ore.: Wipf and Stock, 2008), pp. 100-101. He includes items such as the blind spot in

the eye and the inherent instability of the spine.

[2]Undoubtedly many will ask the inevitable question concerning genetic defects, miscarriages and the variety of other things that can go wrong in this process. If this is the handiwork of God, why can't he get it right? This takes us back into the *why* realm, and those are questions for which we are not given answers. The affirmation that we are urged to make is that we trust God's wisdom, as difficult as our circumstances become. This is what the book of Job teaches, as does Ecclesiastes (note Eccles 7:14).

[3]Some press a distinction between macroevolution (change from one species to another) and microevolution (change within a species), and the distinction is not insignificant. Nevertheless in this discussion I would like to focus on the overall concept of evolution.

[4]See the discussion of the range of usage in John Stek, "What Says the Scripture?" in *Portraits of Creation,* ed. H. J. van Till (Grand Rapids: Eerdmans, 1990), pp. 216-20.

[5]For a discussion of this option and others see D. Young, "The Antiquity and the Unity of the Human Race Revisited," *Christian Scholar's Review* 24, no. 4 (1995): 380-96.

[6]Obviously this issue requires much more in-depth treatment but is outside of the focus of this book, which is focused on Genesis 1, not Genesis 2.

Proposition 17: Resulting Theology in This View of Genesis 1 Is Stronger, Not Weaker

[1]In biblical terms we could point to the four-hundred-year delay in giving the promised land to Abraham's descendants (Gen 15), or even to the long wait for the return of Christ. Daniel 9 also offers an example in the long period of time during which restoration of the people of Israel will occur.

[2]Orson Scott Card, *Prentice Alvin* (New York: Tor, 1989), pp. 260-62, excerpts used by permission.

[3]Some paragraphs of this sections have been taken from John Walton and Andrew Hill, *Old Testament Today: A Journey from Original*

Meaning to Contemporary Significance (Grand Rapids: Zondervan, 2004), p. 129.

Proposition 18: Public Science Education Should Be Neutral Regarding Purpose

[1]For my definition of *empiricism* see note 6 in chapter 13 (p. 164).

[2]Not content with an empirically based methodology, metaphysical naturalism mandates the restriction of *reality* to that which is material.

[3]National Association of Biology Teachers, "Statement on Teaching Evolution," *The American Biology Teacher* 58, no. 1 (1996): 61

[4]Ibid. In the list of what they refer to as "tenets of science, evolution and biology education," the NABT statement read: "The diversity of life on earth is the outcome of evolution: an unsupervised, impersonal, unpredictable and natural process of temporal descent with genetic modification that is affected by natural selection, chance, historical contingencies and changing environments."

[5]*Methodological naturalism* refers to the self-imposed restriction that no appeal will be made to supernatural agency. It accepts the premise that mechanisms themselves are dysteleological without extrapolating those operating principles to the larger metaphysical enterprise.

[6]Revised statement from 2004 can be found at <sci.tech-archive.net/Archive/sci.bio.evolution/2006-01/msg00177.html> or another site <www.natscience.com/Uwe/Forum.aspx/evolution/2103/National-Association-of-Biology-Teachers>.

[7]Evidenced in the statements from the NABT, which describe evolution as an important natural process explained by valid scientific principles. They are anxious to "separate science from non-scientific ways of knowing, including those with a supernatural basis such as creationism. Whether called 'creation science,' 'scientific creationism,' 'intelligent design theory,' 'young earth theory,' or some other synonym, creation beliefs have no place in the science classroom. Explanations employing nonnaturalistic or supernatural events, whether or not explicit reference is made to a supernatural being, are outside the

realm of science and not part of a valid science curriculum" (National Association of Biology Teachers, "Statement on Teaching Evolution," p. 61).

[8]Ibid.

[9]Recognizing that what appears to be irreducibly complex may or may not actually be so.

[10]Advocates of design may be able to claim that it contains no theistic a priori, but no claim of teleological neutrality can be sustained.

[11]On the other hand, metaphysical issues cannot and should not be entirely eliminated. Material ontology and the methodological naturalism associated with empiricism are foundational for science, so those particular metaphysical positions need to be assumed.

[12]The deduction that something is likely to be the result of design or random development is itself a stage of rationalism that is the normal result of empirical science. As such it stands as metaphysically transitionary, with the real metaphysics being engaged only when the discussion moves to the nature of the designer or the absence of one.

[13]"Questions About Science Education Policy," question 3, on the Discovery Institute's website (August 13, 2008) <www.discovery.org/csc/topQuestions.php>.

[14]Ibid.

[15]See Hugh Gauch, *Scientific Method in Practice* (Cambridge: Cambridge University Press, 2002).

[16]Of course it must be recognized that "teleological neutrality" may be an impossibility. At least fairhandedness ought to be expected.

[17]Thomas B. Fowler and Daniel Kuebler, *The Evolution Controversy* (Grand Rapids: Baker Academic, 2007), p. 355 (italics theirs).

Summary and Conclusions

[1]See the important article by Timothy Larsen, "'War Is Over, If You Want It': Beyond the Conflict Between Faith and Science," *Perspectives on Science and Christian Faith* 60, no. 3 (2008): 147-55.

[2]Thomas B. Fowler and Daniel Kuebler, *The Evolution Controversy* (Grand Rapids: Baker Academic, 2007), p. 354.

[3]This despite the inadequacy of natural selection and random mutation to offer comprehensive mechanisms for the type of prolonged change over time evidenced in the fossil record and other places. See details in Fowler and Kuebler, *Evolution Controversy,* chapter 5, helpfully summarized on pp. 346-47 and the table on p. 348.

[4]Gerald Runkle, *Good Thinking,* 2nd ed. (Austin, Tex.: Holt, Rinehart & Winston, 1981), p. 271.

Index

Black Self-Concep[illegible]

Black Self-Concept

Implications for Education and Social Science

Edited by

James A. Banks
University of Washington, Seattle
and
Jean Dresden Grambs
University of Maryland

McGraw-Hill Book Company
New York, St. Louis, San Francisco, Düsseldorf, Johannesburg,
Kuala Lumpur, London, Mexico, Montreal, New Delhi,
Panama, Rio de Janeiro, Singapore, Sydney, Toronto

Black Self-Concept

Library of Congress Catalog Card Number 74-159497

345678910 MUBP 76543

This book was set in Theme by Creative Book Services, division of McGregor & Werner, Incorporated, and printed and bound by Murray Printing Co. The designer was Creative Book Services, division of McGregor & Werner, Incorporated. The editor was William J. Willey. Sally Ellyson supervised production.

03610-1

Contents

The Contributors

James A. Banks is Associate Professor of Education at the University of Washington, Seattle. He received his bachelor's degree in education from Chicago State College, and the master's and doctor's degrees from Michigan State University. Professor Banks taught in the public schools of Joliet, Illinois, and at the Francis W. Parker School in Chicago. He serves as a social studies and urban education consultant to school districts throughout the United States. His articles have appeared in such journals as the *Instructor, College Composition and Communication,* and *Social Education.* He is the author of *March Toward Freedom: A History of Black Americans* and *Teaching The Black Experience: Methods and Materials.* He co-edited and contributed to *Teaching Social Studies to Culturally Different Children* and *Teaching The Language Arts to Culturally Different Children.*

Jean Dresden Grambs is Professor of Education at the University of Maryland. She graduated from Reed College and received her master's degree and doctorate in education from Stanford University. Professor Grambs served as a high school teacher and taught at Stanford prior to her appointment at Maryland. She currently serves as a consultant to many organizations dealing with problems in intergroup education. Professor Grambs has contributed to such journals as *Social Education, Elementary Principal, Educational Leadership,* and *Teachers College Record.* She is the author of *Schools, Scholars and Society; Intergroup Education: Methods and Materials;* and co-author of *Modern Methods in Secondary Education* and *Black Image: Education Copes with Color.*

Cynthia N. Shepard is Lecturer in Education at the University of Massachusetts and Associate Professor of Education at Texas Southern University. She received her bachelor's and master's degrees from Indiana State University and is currently a doctoral candidate at the University of Massachusetts, specializing in International Education. Mrs. Shepard formerly served as an Educational Representative for the IBM Corporation and has taught at Allen University, Indiana State University, and the University of Nairobi (Kenya). She was Director of Teacher Corps at the University of Massachusetts from 1969 to 1971. Mrs. Shepard has contributed to *Trend.*

Donald H. Smith is Professor of Education and Director of Educational Development at the Bernard M. Baruch College of the City University of New York. He graduated from the University of Illinois (A.B.), DePaul University(M.A.), and the University of Wisconsin (Ph.D.). He was a teacher-counselor in the Chicago public schools and was Director of the Center for Inner-City Studies at Northeastern Illinois State College in Chicago. Professor Smith has also served on the staffs of the University of Pittsburgh and the Urban Coalition. He has contributed to such journals as the *Journal of Human Relations,* the *Journal of Negro Education,* and the *Journal of Teacher Education.* He also contributed to the book, *Racial Crisis in American Education.* Professor Smith is a consultant to many organizations dealing with problems in urban education and race relations.

Alvin F. Poussaint is Associate Professor of Psychiatry and Associate Dean of Students at the Harvard Medical School. He received his bachelor's degree from Columbia College and his doctor of medicine from Cornell University Medical College. He earned his master's degree at UCLA. He previously served as the Southern Field Director of the Medical Committee for Human Rights in Jackson, Mississippi, and taught at the Tufts University Medical School. He is a former Director of Psychiatry at Columbia Point Health Center in Boston. Dr. Poussaint has contributed to many publications, including *The New York Times Magazine,* the *Journal of Pediatrics*, and the *American Journal of Psychiatry*. He is currently a contributing editor of *The Black Scholar.*

Carolyn Atkinson is a Lecturer in the Department of Sociology and Fellow at The Urban Center at Columbia University. She received her bachelor's degree at Vassar College and her master's degree at Harvard University, where she is currently completing requirements for her doctorate. Miss Atkinson has done extensive research on the sociology of minority groups, and has served as Associate Project Director of a policy research study on alternative transfer payment systems, and as Project Director of the American Jewish Committee's study of Negro anti-Semitism. Her articles have appeared in journals such as the *Journal of Negro Education,* the *American Behavioral Scientist,* and the *Public Administration Review.* She is the co-author of *Post Secondary Education and the Disadvantaged* and *Sociological Implications of Alternative Income Transfer Systems.*

Bradbury Seasholes is Director of Political Studies at the Lincoln Filene Center for Citizenship and Public Affairs, and Associate Professor of Political Science at Tufts University. Professor Seasholes received his bachelor's degree at Oberlin College, and both the master's and doctor's degrees from the University of North Carolina. Prior to his present appointment, he taught at the Massachusetts Institute of Technology. Professor Seasholes is a specialist in political participation and urban problems. At the Lincoln Filene Center, he is responsible for the Center's research and evaluation in the field of political behavior; the development of political attitudes and orientation; urban politics; and political education. He has contributed to several books and edited *Public Education in Massachusetts; The Public Servant; Voting, Interest Groups and Parties;* and *Law and Order.*

Nancy L. Arnez is Professor and Director of the Center For Inner-City Studies at Northeastern Illinois State College. Professor Arnez earned her bachelor's degree at Morgan State College, and the master's and doctor's degrees from Teachers College, Columbia University. Professor Arnez taught in the Baltimore (Maryland) public schools and served as Director of Student Teaching at Morgan State College prior to her present appointment. Listed in *Who's Who of American Women,* Professor Arnez is a poet as well as a professional educator. She has contributed articles to the *English Journal,* the *Journal of Teacher Education,* and the *Journal of Negro Education.* She is the author of *Booknotes to Moll Flanders.* Dr. Arnez's poetry has been published in a wide variety of magazines and in her book of poetry, *The Rocks Cry Out.*

James A. Goodman is Vice-Provost for Minority Affairs and Associate Professor in the School of Social Work and the Department of Sociology at the University of Washington. Professor Goodman received his bachelor's degree from Morehouse College, his master's degree in social work at Atlanta University, and his doctorate from the University of Minnesota. Dr. Goodman served as a clinical social worker in the Los Angeles Health Department, as Chief Social Worker at the Alcoholic Clinic (Los Angeles), and Director of Social Services for the Los Angeles City Health Departments. He taught at the California College of Medicine and at the Universities of Southern California and Minnesota. Professor Goodman has also served as Acting Director of Black Studies at the University of Washington. His articles have appeared in journals such as *California Health* and *The Black Politician.*

Barbara A. Sizemore is Director-District Superintendent of the Woodlawn Experimental Schools District Project in Chicago. She is also an Instructor at the Center for Inner-City Studies at Northeastern Illinois State College. Mrs. Sizemore received her bachelor's and master's degrees at Northwestern University and is a doctoral candidate at the University of Chicago, where she is specializing in educational administration and urban affairs. She taught in the Chicago public schools for many years and was principal of the Forrestville High School. She was a Staff Associate at the Midwest Administration Center of the University of Chicago from 1967 to 1969. Mrs. Sizemore's articles have appeared in *Educational Leadership* and the *Notre Dame Journal of Education.* She contributed to the book, *Racial Crisis in American Education.* She serves frequently as a national consultant on problems in urban education.

Introduction

In September 1963, a conference was held at the Lincoln Filene Center at Tufts University to explore the various dimensions of black self-concept, to delineate ways in which the school could enhance the self-images of black children, and thereby increase their academic achievement and emotional growth. The position papers and discussions which grew out of this conference were published in a widely circulated and quoted book, *Negro Self-Concept* (McGraw-Hill, 1965). The calling of this conference and the publication of the book which presented the position papers and conference discussions were significant events. They indicated that educators in the mid-1960s were not only becoming increasingly sensitive to the unique problems which black children experience in our schools and society, but were searching for creative ways to solve them. The reception which *Negro Self-Concept* received, and the dialogue which it stimulated, suggested that the topic was of interest to educators. The book was widely read and chapters were reprinted in a number of anthologies.

Since the publication of *Negro Self-Concept* in

1965, a number of notable changes have occurred within our society and within black communities which suggest the need for a renewed dialogue among educators and policy makers about the role of education in shaping the view black children have of themselves. There is a strong movement among blacks to reject their old identity, shaped to a large extent by white society, and to create a new one, shaped by blacks. Many blacks abhor the word Negro, wear Afro hair styles and clothing, and learn African history. Blacks are trying to gain control of their schools and political control of their communities. The call for Black Power and the slogan Black is Beautiful are rallying cries of this movement. The search for a unique black identity has resulted in many unprecedented changes within the black community.

While many blacks, particularly youth, sought a new identity during the 1960s, a critical reaction against black demands emerged from the power circles of the white community. A cry for "law and order" by influential white leaders during the last years of the sixties was symptomatic of latent anti-black forces coupled with a rising hostility to youth demands for a withdrawal from Vietnam. When the 1970s opened, the black American found himself in a curious predicament. He was without strong national leadership, and the civil-rights progress of the 1960s seemed to be losing momentum. The atmosphere throughout the nation was ominous. Blacks were still the last hired and the first fired. Their children attended the worst schools, and efforts for community control of black schools had been largely unsuccessful. Many teachers, both black and white, believed that black youth could not learn on a par with white children. All of these events had an impact on the evolution of black identity.

We feel that the changes which have taken place in our society and within black communities since the publication of *Negro Self-Concept*, and the urgent need for the school to help the black child clarify his identity, warrant renewed exchange among educators and policy makers about the relationship between education, social science, and the development of the black child and youth. This book is designed to stimulate such exchange.

Although the black revolt of the 1960s gave rise to a new search for identity among many blacks, it is extremely difficult to make empirically sound statements about the impact which the black revolt has had on the self-concepts and attitudes of black children and youth.

Prior to the 1960s, studies by such researchers as the Clarks, Goodman, Trager and Yarrow, and Hartley[1] consistently concluded that most black children inculcate the negative evaluations of their race which are perpetuated by the larger white society, and thus develop lower self-concepts than white children. This research also indicated that blacks evidence more incorrect racial self-identifications and conflict when they are asked to express racial preferences. Since most of this research was conducted prior to the black revolt of the past decade, it is important to consider whether the black revolt has significantly affected the black self-concept in a positive direction.

While recent research is inconclusive and somewhat contradictory, the *bulk* of it indicates that the black revolt *has not* significantly changed the self-concepts and self-evaluations of *most* black children and youth. In summarizing the available research Goldschmid (1970) writes, "It is not surprising, then . . . that many black people incorporate negative views of their own subculture and identify with white middle-class standards. This may be particularly true of upwardly mobile Negroes. . . ."[2]

Recently a number of writers have rejected the *assumptions* on which most "black self-concept" research is based, have questioned its research designs and validity, and argue that it does not accurately describe the self-esteem of today's black youth. In a new book, E. Earl Baughman argues that we need new assumptions to guide black self-concept research, and posits an alternative hypothesis which states that ". . . The supply of self-esteem is not, and has not been, less for the black than the white."[3] He contends that the black child's positive self-concept is formed *prior* to his interactions with the white world. Jean D. Grambs, in Chapter 8 of this book, reacts to Baughman's assumptions and hypothesis. Baughman's argument is based on a hypothesis which is largely unverified, although he does cite two recent studies which support his point of view.

It is clear from the research reviewed by Banks in Chapter 1, and from the studies discussed by Baughman, that much careful, intensive research must be undertaken before we can make conclusive statements about the status of the black self-concept today. Part of the problem results from the fact that researchers have not reached consensus about what *self-concept* is. Although many definitions of the concept have been offered, such as the recent one by Gergen,[4] different researchers often define the concept differently, and therefore measure different kinds of behavior when they are supposedly studying the same concept. Some researchers, such as Brookover and his associates, have attempted to delineate a *specific* kind of self-concept, such as the *self-concept of ability.*[5] Since an individual may have many different kinds of self-concepts, it may be that we will make greater strides in self-concept research when more efforts are made to study *specific* kinds of self-concepts rather than a *generic* or *global* kind, such as most researchers have studied.

More research and standardized definitions of *self-concept* are needed before we can make conclusive statements about black self-esteem. However, the current research on a *global* type of self-concept and our own observations and analyses force us (the editors) to conclude that the positive impact of the black revolt of the last decade has been exaggerated by many writers, and that the perceptions and attitudes which many black children have toward themselves are still negative, despite the profoundly positive affects of the black revolt on the black psyche. A vocal black minority shouts Black Power and Black is Beautiful, but most black children still live within a world where to be black produces feelings of shame, despair, and anger. These children find it difficult to believe that they are beautiful when so many of the conditions within their home, school, and community do not support such a belief. While this is our point of view, we respect the opinions of writers such as Baughman, and believe that they merit careful study and testing.

Some of the contributors to this volume do not accept our conclusions. This book contains papers which present a wide range of views regarding the status of the black self-concept today. In her paper, Nancy L. Arnez rejects the "negative black self-concept" hypothesis; Barbara A. Sizemore examines the problem from a fresh perspective. Authors with *differing* opinions, not only about the status of the black self-concept today, but also about the role of educators in shaping the positive development of black children were invited to write chapters. Since the research is somewhat contradictory and inconclusive, and clear-cut answers do not exist, we felt that discussions by authorities of different opinions and races would stimulate fruitful dialogue among educators and policy makers. This exchange might result in useful hypotheses for further research and in the development of guidelines for sound educational programs for black children and youth.

Each of the contributing authors was asked to approach the problem of education and the black self-concept from his own perspective. All of the chapters are unified by the authors' concern with creating that school environment which can facilitate the attainment of positive and healthy self-concepts by black children. The phenomena referred to by all authors is a *global* or *generic* set of attitudes and perceptions which an individual has toward himself that significantly influence his academic and social behavior. Most authors use the term *self-concept;* others prefer *identity.* All of the contributors agree that an individual's self-concept is shaped by his social interactions with the persons in his environment; it is not inborn. While the bulk of the book consists of unpublished papers written especially for this volume, several selections which we felt were especially pertinent to the topic are reprinted from other sources.

The selection by Cynthia N. Shepard, which constitutes the *Prologue,* sensitively describes how the world in which her black son Mark lives caused him to reject blackness and to value whiteness. The problem which this essay illuminates is central to the chapters which follow. Mark's social-psychological environment affects many other black children in a similar way, as several of the contributors indicate.

Arguing that white racism is the major cause of the black child's low self-image, James A. Banks suggests in Chapter 1 that we must initiate a direct and unprecedented attack on white racism and create new "significant others" for black children in order to increase the black child's self-concept. He contends that whites and blacks have somewhat separate roles to play in augmenting the black child's self-image. Banks's chapter stresses the role of whites; Chapter 5 by Nancy L. Arnez speaks more directly to the role which blacks must play. The research which the author reviews provides useful, if nonconclusive, guidelines for structuring and implementing intervention programs to reverse negative racial attitudes and self-concepts.

Donald H. Smith, in Chapter 2, provides a perceptive and stimulating discussion of the black revolt of the 1960s, and illuminates how the revolt has permeated the American school. After detailing how white racism has "poisoned the American school," he makes some proposals for change which he feels are imperative if the school is to help black children attain more positive self-perceptions and higher academic achievement. Smith's paper is polemic and will undoubtedly provoke discussion.

Chapter 3, by Poussaint and Atkinson, raises questions about the relationship among variables which affect the motivation of black youth. Factors related to motivation such as self-concept, patterned needs, and the rewards offered by society are analyzed. While the authors regard the self-concept as a cogent factor which affects the motivation and achievement of black youth, they suggest that it might "be a weaker motivator of behavior than the motive of self-assertion and aggression." The authors' analysis of variables influencing the motivation of black youth extends and clarifies points raised in earlier chapters.

Poussaint and Atkinson argue that blacks must attain a sense of control over their environment in order to gain self-confidence. Political power will enable oppressed ethnic groups to control their own destinies. In Chapter 4, Bradbury Seasholes discusses the political socialization of black youth, and explores the relationship between political effectiveness, self-concept, and personal stability. He suggests that the school can best help black children attain a sense of political efficacy by providing opportunities for them to devise political strategies which can influence the making of public policy.

Authors of the previous chapters contend that many black children evidence low self-esteem and negative self-concepts. Nancy L. Arnez, writing in Chapter 5, rejects this postulate, and argues that writers who argue that blacks have negative self-concepts are relying on research conducted before the advent of the black revolt of the 1960s. Arnez reviews research which supports her argument. The author argues that it is demeaning to continue to write about the negative self-image of black children. Arnez suggests that teachers must recognize the black child's already positive self-concept and create teaching strategies to enhance it. She describes some promising instructional techniques, using the new black literature as the primary teaching material. Such literature, the author contends, can reinforce the black child's positive self-image. Many of Arnez's conclusions contradict conclusions derived by other contributors.

In Chapter 6, James A. Goodman discusses the effects of racism on the development of black identity. He uses the conceptualizations of identity theorists such as Erikson and McCandless, and insights derived from learning theory, to explain the individual and group factors which influence the development of black self-esteem. Goodman reveals ways in which the public school has damaged the identities of black children, and makes some sound proposals regarding ways in which the school can enhance the self-images of black children and youth. He suggests the need for new assumptions to guide educational planning and instruction.

In Chapter 7, Barbara A. Sizemore raises the powerful question: "Why does the black man feel that social science has failed him?" She then provides a review of the problems of social science research, suggests a new theoretical framework for social science, and critically assesses several controversial studies of black Americans, including the research of Jensen, Moynihan, and Banfield. Sizemore discusses the relationship between social science, education, and the black identity. The author presents her views regarding the proper components of education for a positive black identity.

In the final chapter, Jean Dresden Grambs attempts to assess the changes which have resulted from the black revolt of the 1960s, and to describe their impact on the self-concepts and behavior of black youth. Professor Grambs also tries to delineate the appropriate responses which educators should make to the changes resulting from the black revolt of the last decade. The author re-analyzes the self-concept construct which she discussed in *Negro Self-Concept* (1965). She concludes that a redefinition of the concept is needed, and that the notion of *competence* and *achievement motivation* must loom large in any attempt to define self-concept for educators.

It is our hope that this book of essays will stimulate continuing examination by educators about the school's role in enhancing the self-perceptions and identities of black youth. We are grateful to the authors who wrote chapters in response to our requests, to Bradbury Seasholes for revising his chapter which appeared in this book's predecessor, and to the authors and publishers who permitted us to reprint the previously published materials. We extend a very special thanks to Cherry McGee Banks for preparing the Index. She worked tirelessly to complete this tedious but important task.

James A. Banks
Jean Dresden Grambs

NOTES

1. Kenneth B. Clark and Mamie P. Clark, "Skin Color As a Factor in Racial Identification of Negro Pre-School Children," *Journal of Social Psychology* 11 (1940), pp. 159-167; Mary Ellen Goodman, "Evidence Concerning the Genesis of Interracial Attitudes," *American Anthropologist* 48 (1946), pp. 624-630; Helen G. Trager and Marian Radke Yarrow, *They Learn What They Live* (New York: Harper, 1952); E. L. Hartley, *Problems in Prejudice* (New York: Crown, 1946).
2. Marcel L. Goldschmid, ed., *Black Americans and White Racism: Theory and Research* (New York: Holt, 1970), p. 19.
3. E. Earl Baughman, *Black Americans: A Psychological Analysis* (New York: Academic, 1971), p. 42.
4. Kenneth J. Gergen, *The Concept of Self* (New York: Holt, 1971). Gergen defines self-concept as follows: "The notion of self can be defined first as process and then as structure. On the former level we shall be concerned with *that process by which the person conceptualizes (or categorizes) his behavior—both his external conduct and his internal states.* On a structural level, our concern is with *the system of concepts available to the person in attempting to define himself.*" Pp. 22-23.
5. Wilbur B. Brookover, Edsel Erickson, and Lee Joiner, "Self-Concept of Ability and School Achievement," in James A. Banks and William W. Joyce, eds., *Teaching Social Studies to Culturally Different Children* (Reading, Mass.: Addison-Wesley, 1971), pp. 105-109.

Acknowledgments

The editors gratefully acknowledge the cooperation of the following authors and publishers who granted us permission to use their materials:

Cynthia N. Shepard, "The World Through Mark's Eyes," *Saturday Review* (January 18, 1969), p. 61, Copyright © 1969, by Saturday Review, Inc. Reprinted with permission of the author and publisher. Donald H. Smith, "The Black Revolution and Education," in Robert L. Green (Editor), *Racial Crisis in American Education* (Chicago: Follett Educational Corporation), pp. 57-71, excluding footnotes 1, 6, 10, and 12. Copyright © 1969, by Follett Educational Corporation. Reprinted with the publisher's permission. Alvin Poussaint and Carolyn Atkinson, "Black Youth and Motivation," *The Black Scholar,* vol. 9 (March 1970), pp. 43-51, Copyright © 1970 by The Black World Foundation. Reprinted with permission of the publisher and Alvin Poussaint. Lines from *What Shall I Tell My Children Who Are Black* by Margaret Burroughs. Reprinted with the author's permission. A brief selection from Dick Gregory (with Robert Lipsyte), *Nigger: An Autobiography*

(New York: E. P. Dutton and Company, Inc., 1964, pp. 45-46, Copyright © 1964 by the E. P. Dutton and Company, Inc. Reprinted with permission of the publisher. World Rights to use the selection from *Nigger* was granted by George Allen and Unwin Ltd. Book Publishers, London. Quotes from Orde Coombs, "Books Noted," *Negro Digest*, vol. 18 (November 1968), pp. 74 and 75, Copyright © 1968 by *Negro Digest.* A quote from Don L. Lee, "Black Poetry," *Negro Digest,* vol. 17 (September/October 1968), p. 32, Copyright © 1968 by *Negro Digest*. Reprinted with the permission of the authors and *Black World.* Three quotes from Allen Wheelis, *The Quest For Identity* (New York: W. W. Norton and Company, 1968), pp. 179, 182, and 200, Copyright © 1968 by the W. W. Norton and Company. Reprinted with the publisher's permission. Three quotes from Helen G. Trager and Marian R. Yarrow, *They Learn What They Live,* (New York: Harper and Brothers, 1952), pp. 155, 150, and 341, Copyright © 1952 by Harper and Brothers. Reprinted with permission from Harper & Row Publishers. A quote from Charles Keil, *Urban Blues* (Chicago: University of Chicago Press, 1967), p. 27, Copyright © 1967 by the University of Chicago Press. Reprinted with permission from the publisher. A quote from Dante P. Ciochetti, "CBE Interviews: Kenneth B. Clark," *Bulletin of the Council for Basic Education,* 14 (November 1969), pp. 15-16, Copyright © 1969 by the Council for Basic Education. Reprinted with the publisher's permission. A quote from Eleanor B. Leacock, *Teaching And Learning in City Schools: A Comparative Study,* (New York: Basic Books, Inc., Publishers, 1969), p. 203, Copyright © 1969 by Basic Books, Inc. Reprinted with the publisher's permission.

Prologue

The World through Mark's Eyes

Cynthia N. Shepard

I would like you to know my son Mark, who is now five years old. Although he has not yet attended kindergarten, he can both read and write, and can accurately identify colors and forms with an acuity beyond his years. He collects American flags, and pictures and ceramics of our national emblem, the eagle. He learned from somewhere on his own initiative the Pledge of Allegiance, which he recites with deep fervor. He only asked me the definitions of those difficult words: *indivisible, liberty, justice.* My precious, precocious Mark is very proud of his white, Anglo-Saxon heritage. But, he's black: a beautifully carved and polished piece of black American earth.

You may debate with me whether I should have taught him from birth that he is black. Instead, I invite you to see the world through Mark's eyes. Mark learned to read when he was three years old—books based on the white American style of life with pictures of blond, blue-eyed suburbia, with decent interspersing of browns and brunets—but no blacks. He watched the "educational" newsreels on television, which for him reinforced the rightness of whiteness. The man in the white hat—beating the black man with a billy club and then kicking him into insensibility—was the good guy. He was the protector of our individual rights. The books said so.

Black is the night which Mark fears, vanquished by the white of day. White is the knight on the white horse charging the black stains of daily living, and they all vanish. Black is unwanted; black is weak and easily defeated; black is bad.

I took Mark South with me and placed him in an all-black nursery school while I taught during the day. The first evening, when I brought him home, he was in tears, writhing and retching in painful confusion. "Why did you make me go to school with all those Negroes?"

Then, just like NOW, I dig! Intellectuality had blocked my insight, creating of me a blind broad and of my black son a white racist. In his innocence—or highest sophistry, you see—he had intuitively perceived race not as a color, but as an attitude that he did not exemplify. My arguments to the contrary were completely hushed by his own words: "You said I could be anything I choose, and I choose to be white. I am white."

I returned North and searched both public and university libraries for literature with both pictures and narrative with which he might relate. *Little Black Sambo?* Oh no, dear God! Where are the black men of history, the Nat Turners, the Veseys, the Prossers? The uncompromising, unprecedented, unheralded warriors for true democracy?

I found a book about John Henry, with all the usual legendary verbiage. But it had pictures—pictures of John Henry as a big, black and beautiful baby; pictures of a handsome, adventurous black youth; and then, a picture of a dynamic, virile, muscle-bound black man. John Henry, the steel-driving man: a beautiful portrayal of black maleness, bared to the waist, swinging that hammer with all his might. It is with *that* picture that my son finally identified: an uncompromising image of black masculinity. That's what it's all about, baby.

I doubled my search for books that pictured black and white children running and laughing together, while black and white mothers shopped and lunched together, while black and white fathers worked and played together. I found a few. Mark had no difficulty identifying me in the pictures, but only recently could he find himself.

Eventually, I overheard him speak of himself as a little brown boy, and I rejoiced—deeply, I say—that he was finding his way out.

Now, I have brought him East and have enrolled him in a kindergarten where all the other children are white. But I have not yet been able to send him to that school, although soon legally I will have to send him. What can be done to save my child from a plunge into utter confusion? What can be done to help my little black boy?

What can the world of education do to alleviate his pain—and mine? Must he grow like Topsy: confused, angry, alienated, lighting chaotic fires from the burning bitterness within? By America's guilt-ridden permissiveness, will he be ignored to become a black-helmeted, black-booted, black-bigoted replica of the swastika? Will my son see the necessity of asserting his blackness, his maleness, militantly and insensitively, riding roughshod over all who might in any manner oppose? Or, can the world of education, with all its demonstrated expertise, utilize the precociousness of my little black boy for the building of a better world for all people? How? When? *Now* is the answer.

Today Mark told me he is going to be an eagle and fly high above the earth where nobody will be able to stop him. That speaks to me. I gaze into Mark's face as he lies peacefully sleeping. With all the normal pains of growing, what utter, needless trauma I know he must also face tomorrow—unless change is *made* to happen.

Bitterness wells up within me, too, and I wretchedly whisper into his unheeding ear the words of a writer who must also have known deep human agony over the inhumanity of man to man:

O pardon me, thou bleeding piece of earth,
That I am meek and gentle with these butchers.

What shall I tell my children who are black
Of what it means to be a captive in this dark skin?
What shall I tell my dear one, fruit of my womb,
Of how beautiful they are when everywhere they turn
They are faced with abhorence of everything that is black.
The night is black and so is the bogeyman.
Villains are black with black hearts.
A black cow gives no milk. A black hen lays no eggs.
Bad news come bordered in black, mourning clothes black,
Storm clouds, black, black is evil
And evil is black and devils food is black.[1]

Margaret Burroughs

1

Racial Prejudice and the Black Self-Concept

James A. Banks

After a long history of racial exploitation, violence, and oppression, our nation has finally acknowledged the fact that American society is afflicted by a chronic and agonizing malady—white racism. It took several hundred lives—most of them black—the burning of millions of dollars worth of property, and an expensive commission report to actualize this rude awakening. Wrote the National Advisory Commission on Civil Disorders:

> *This is our basic conclusion. Our nation is moving toward two societies, one white, one black—separate and unequal. . . . the most fundamental [cause of the riots] is the racial attitude and behavior of white Americans toward black Americans. . . . Race prejudice has shaped our history decisively; it now threatens to affect our future. . . . White racism is essentially responsible for the explosive mixture which has been accumulating in our cities since the end of World War II.*[2]

Our painful recognition of the racial crisis which confronts us, and the black man's irrevocable determination to share the "American Dream" have profoundly affected the attitudes of most Americans and increased racial polarization throughout the nation. More than any of his predecessors, the young black American today is determined to achieve a sense of dignity in American society, and to create and perpetuate a new and positive identity. The flames that burned in Watts, the blood that ran in Detroit, and the willingness of black leaders to chance assassination by taking strong public stands on social issues indicate that many black Americans are willing to pay almost any price to secure those rights which they believe are theirs by birthright.

The reactions by the white community to the black man's new militancy have been intense and extreme. A "law and order" cult has emerged to stem the tide of the black revolt. Promises to bring law and order to the streets guarantee public officials election victories, yet few constructive actions have been taken by national and local leaders to eliminate the hopelessness, alienation, and poverty which breed crime and violence. Although the law and order movement is directed primarily toward the poor, the black, and the alienated (such as the Washington, D.C. "no-knock law"), the most costly and destructive crimes in this nation are committed by powerful syndicates, corrupted governmental officials, and industries that pollute our environment, and not by the ghetto looter and petty thief. The public official who swindles thousands of dollars of Title III funds commits a much greater crime than the looter who takes a television set in the heat of a ghetto rebellion.

The black man's status in relationship to whites has actually worsened in recent years. While the gap between whites and nonwhites is narrower in educational achievement than it was a decade ago, the gap in all other major areas—income, employment, health, and housing—has increased in the last ten years,[3] despite the symbolic and legal gains which accrued from the black revolt of the 1960s. Public schools in the South and in most major cities are increasingly becoming segregated. We are rapidly returning to "two separate and unequal societies," which presumably the Supreme Court decision of 1954 had declared obsolete.

In modern American society we acquire identity from other human beings ("significant others"), and incorporate it within ourselves. A person in our society validates his identity through the evaluations of "significant others." However, the average black American has never been able to establish social or self-identity that is comparable in terms of social valuation to that of the white majority. The ideal self in America has been made synonymous with Caucasians, and particularly middle-class whites. In his quest for identity, the black man has begun to ask, "Who am I in relation to other races and ethnic groups?" This heightened awareness has caused many black Americans to reject their old identity, which was shaped largely by the white community, including the term "Negro," and to search for elements out of which a new identity may be formed. Such elements include a closer identity with African states, intensified racial pride and cohesiveness, and a search for power.

In his attempt to shape a new identity and to gain control over those factors which most profoundly affect his life and destiny, the black man confronts an immense hurdle—a system of white institutional racism which perpetuates racial myths to justify the oppression of the black man. Even though a vocal group of recognized black leaders are telling black youth that black is beautiful, the larger white society repeats the myth that black is evil, bad, and ugly. Indeed, the dilapidated buildings in which they often live, the squalor in their neighborhoods, their inferior schools, and hostile white and Negro teachers all tell black children everyday that black is less desirable than white. No amount of peaching about the achievements of Crispus Attucks and Daniel Hale Williams in social studies classes can successfully counteract these cogent and negative lessons. Writes Kenneth Clark:

> *Children who are consistently rejected understandably begin to question and doubt whether they, their family, and their group really deserve no more respect from the larger society than they receive.*[4]

To help the black child bolster his self-concept, a number of school districts throughout the nation, responding largely to pressure from community groups, have implemented courses in black history, hired more black administrators and teachers, and purchased a flood of new textbooks and multimedia kits which are sprinkled with "selected" black heroes and white characters colored brown. While these efforts,

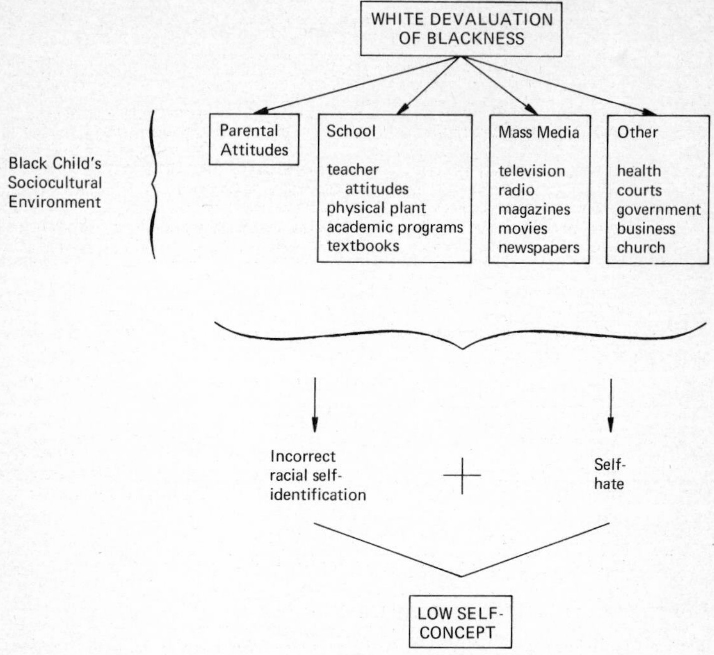

Figure 1. Social determinants of the black child's self-concept.

and many more, are certainly needed to help the black child augment his self-concept, they do little to solve our basic problem. To confront this problem head-on, we must initiate a direct and unprecedented attack on white racism.

Since whites are the dominant and "significant others" in American society, and black children derive their conceptions of themselves largely from white society and its institutions (see Figure I), we are not going to progress significantly in augmenting the black child's self-concept until we *either* change the racial attitudes and perceptions of white Americans or create new "significant others" for black children. However, we should take a multiple approach to the problem. Both blacks and whites have essential but somewhat separate roles to play in enhancing the black child's identity. Blacks should work to create new "significant others" for black children, and provide them with successful models with whom they can identify. In her chapter in this book, Professor Nancy A. Arnez discusses the role of blacks in enhancing the black child's self-image.

However, racism is the main cause of the black child's deflated self-concept, and whites must assume the main responsibility for eliminating racism in this nation and for saving it from destructive polarization. The black child's self-image will become more positive and productive as a byproduct of the elimination of white racism. *This chapter stresses the role of whites in the reconstruction of American society and in increasing the black child's concept of himself.* The author *does not* believe that blacks should remain passive and wait for whites to change their racial attitudes—this may never happen. However, the research reviewed in this paper does suggest that it's going to be difficult, if not impossible, to help black children think highly of themselves as long as they are socialized in a racist society that devalues them and destroys their hearts and minds.

Since our major social problems stem from the negative attitudes which whites have toward blacks, we must drastically modify the attitudes of whites if we are going to make any headway in solving our most urgent and critical problems. Green justifiably criticized the National Advisory Commission on Civil Disorders for failing to suggest ways to deal with white racism after declaring it the cause of the most pressing problems which confront us. Green wrote:

> *The major limitation of the Report is its failure to make recommendations regarding education of the white community. If white racism, as reflected in welfare practices, housing, advertising, mass media, police-community relations, and public school education, is the basic cause of the major problems confronting the black community, the Report is incomplete without major recommendations for programs to assist the white community in eradicating white racism. . . . Sociologists, psychologists, educators, and anthropologists who have become experts on the black community should immediately focus attention on the white community. . . . to design action programs that can be evaluated to determine how we can most effectively and rapidly assist the white community in eradicating racism directed at the black community.*[5]

To illuminate the adverse affects of white racism on the black child's self-concept, research on racial attitudes will be reviewed. The negative attitudes which whites exemplify toward blacks and which blacks often express toward blackness will be discussed. This research review will reveal how the black child's conception of himself and his lowered self-esteem are derived from whites. Research on changing racial attitudes will be presented because it is argued that changed attitudes among whites can enhance the black child's self-image.

CHILDREN'S RACIAL ATTITUDES

Lasker's (1929) pioneering research on children's racial attitudes indicates that young children are aware of racial differences, and illuminates some of the emotional components which accompany racial prejudice.[6] An early study by Minard (1931)also suggests that children's racial attitudes are formed during the earliest years of life.[7] Since the seminal research by Lasker and Minard, a number of other researchers have studied race awareness and racial self-identifications of young children. This research has generally confirmed Lasker's early findings. Horowitz (1939) found that most of the nursery school children which she studied were able to correctly identify race.[8] The work by Goodman (1946) confirmed Horowitz's findings. However, Goodman noted, as the Clarks and Morland were to later confirm, that white children make more correct racial self-identifications than black children.[9] Goodman interpreted the black children's frequent inaccurate racial self-identifications as a manifestation of their unwillingness to accept the fact that they were black and the negative connotations of blackness in American society. Writes Goodman:

> *. . . it is possible . . . that the relative inaccuracy of Negro identifications reflects not simple ignorance of self, but unwillingness or psychological inability to identify with [black] because the child wants to look [white].*[10]

The black children in Goodman's study, unlike the white subjects, manifested "uneasiness, tension, and evasion" when they were asked to make racial self-identifications. Most of the children preferred white to brown dolls when asked to indicate a preference.

The Clarks (1947) also found that nursery school black children were aware of racial differences, but were often unable to make correct racial self-identifications.[11] Most of them preferred white to colored dolls and felt that white dolls were "nice dolls" and that colored dolls "looked bad." Like Goodman, the Clarks found that many of their subjects exemplified strong negative emotions when asked to make racial self-identifications. The black children found it difficult to accept the meaning of blackness in American life. Research by Trager and Yarrow (1952) indicated that children in kindergarten, first, and second grades are acutely aware of racial differences and their social implications.[12] When shown a picture of a group of white children playing with a black boy standing aside from the group, almost all of the children suggested that the black boy was not playing with the other children because he was racially different. In explaining why he thought that the black child was not playing with the other children, one second grade white boy said:

> *Because he is colored and don't want to play with white boys, and white people don't want to play with him because they know he cheats and is too tough.*[13]

The white children in the Trager-Yarrow study were also more able to confront racial differences. More white than black children mentioned race when asked to comment on the picture. The inference made was that black children found the topic of race too painful to discuss freely. As in the Goodman study, many of the children made role assignments which reflected the pervasive stereotypes of blacks and whites when they were given brown and white dolls and asked to assign them roles. A majority of them assigned white dolls to the "good house" and brown dolls to the "poor house." Many of the children said that the colored dolls were maids and that the white dolls were their bosses. As in the other studies, most of the children preferred white to colored dolls. Trager and Yarrow summarize their findings:

> *The findings of this study emphasize the complex learning in the young children's reactions to race. Their concepts and feelings about race frequently include adult distinctions of status, ability, character, occupations, and economic circumstances which may become justifications for further discrimination. For the Negro children these distinctions play an important role in self-other assessment. The factor of race is inextricably related to many problem areas in their lives.*[14]

More recent research by Morland (1963) indicates that children's racial attitudes have not changed substantially since the research conducted by Goodman, the Clarks, and Trager and Yarrow. Morland has studied race awareness and the racial attitudes of Southern black and white children, most of whom attended segregated schools. In several of his studies, he presented subjects with pictures of black and white children and asked them such questions as, "Which looks like your mother?" "Which child would you rather be?" and "Which child do you like?" In one of his studies, a majority of the black children said that they looked more like the white child than the black child. However, 71 percent of the white children indicated that they looked more like the white picture. Sixty-six percent of the black children said that they would rather be the white child; less than 19 percent of the white children said that they preferred being black.[15]

Like Goodman and the Clarks, Morland found that white children were able to make more correct racial self-identifications than black children. In a later study, Morland (1966) also found that black children make more incorrect racial self-identifications.[16] Morland's research also indicates that Southern black children are the most likely group to reject members of their own race.

The research discussed above indicates that both black and white children are aware of racial differences at an early age, and that they prefer white to black when they become aware of the evaluations of black and white in the larger society; their attitudes thus reflect those which are pervasive in our culture. White children make more correct racial self-identifications than black children because black youth frequently find it emotionally difficult to accept the meaning of their blackness and thus difficult to develop positive self-concepts.

ADULTS' RACIAL ATTITUDES

Children are not born with racial antipathies. These attitudes are learned. Children learn them from the adults in their environment early in life. Parents, teachers, and other "significant others" in the child's environment shape his racial attitudes. Trager and Yarrow suggest that children conform to social norms and thus inculcate adults' racial attitudes because of their need for acceptance.

> *His [the child] needs for acceptance by persons important to him require his conformity to their patterns of behavior and attitudes. Deviation from the patterns of behavior and beliefs of his family and group may result in frustration of his needs to be accepted and to belong. Attitudes toward self and toward others are acquired in this process of attempting to secure need satisfaction and to obtain meaning from the confusion of stimuli affecting the individual.*[17]

Since children must conform to the behavior patterns esteemed and exemplified by parents and teachers in order to avoid frustration of their needs for acceptance, we will not be successful in changing children's racial feelings until we can modify the attitudes of the significant adults in their environment. The school is obviously limited in what it can do to change parents' racial attitudes, although attempts should be made to involve parents in intergroup education programs. If the school is to formulate and implement intergroup education programs that make a difference, parents must be actively involved in the earliest stages of planning. Educators must play an active and aggressive role in communicating to parents the urgent need to mitigate the racial crisis in the nation, and seek bold and innovative ways to solicit parent participation in attempts to mitigate racial antipathies.

Research suggests that teachers, next to parents, are the most significant others in children lives, and that classroom teachers play an important role in the formation of children's attitudes and predispositions.[18] A study by Davidson and Lang (1960) indicates that the assessment that a child makes of himself is related to the assessment "significant people" make of him.[19] The study demonstrated that a child's self-appraisal is significantly related to his perception of his teacher's feelings. Further, the more positive the child's perception of his teacher's feeling, the better was his academic achievement and classroom behavior as rated by his teachers.

In both subtle and overt ways, teachers influence their students' racial feelings and self-perceptions. The teacher who believes in the superiority of the white race communicates this belief to children by both his actions and verbal behavior. Recalling his painful school experiences in his autobiography, *Nigger,* Dick Gregory relates the following incident. The teacher had asked each child in the class except Gregory how much his father would give to the Community Chest. Fatherless Gregory, longing for a father, decided to "buy" himself a Daddy and pledge money that he had made shining shoes and selling papers:

> *I stood up and raised my hand.*
> *"What is it now?"*
> *"You forgot me."*
> *She turned toward the blackboard. "I don't have time to be playing with you, Richard."*
> *"My Daddy said he'd . . . "*
> *"Sit down, Richard, you're disturbing the class."*
> *"My Daddy said he'd give . . . fifteen dollars."*
> *She turned around and looked mad. "We are collecting this money for you and your kind, Richard Gregory. If your Daddy can give you fifteen dollars you have no business being on relief."*
> *"I got it right now, I got it right now, my Daddy gave it to me to turn in today, my Daddy said . . . "*
> *"And furthermore," she said, looking right at me, her nostrils getting big and her lips getting thin and her eyes opening wide, "we know you don't have a Daddy."*
> *Helene Tucker turned around, her eyes full of tears. She felt sorry for me. Then I couldn't see her too well because I was crying, too.*
> *"Sit down, Richard."*
> *And I always thought the teacher kind of liked me. . . .*
>
> *I walked out of school that day, and for a long time I didn't go back very often. There was shame there.*[20]

A significant body of research, some of which is reviewed below, suggests that most white American adults harbor negative racial attitudes toward black and other ethnic groups. In most of this research, college students were used as subjects. A number of studies also suggest that teachers' racial attitudes are similar to those of most Americans.[21]

Bogardus (1925) conducted some of the pioneering research on the racial attitudes of adults. He developed a scale to measure social distance; a technique which was refined and widely used by other researchers. In a study he reported in 1925, college students ranked Turks and Negroes below all other racial and ethnic groups.[22] In an attempt to find out how students developed racial antipathies, he asked the respondents to describe situations which caused their negative feelings. The following account is especially revealing because it is more of a rationalization for prejudice than an explanation of its origin:

> *My first encounter with the Negro was in Louisville, Kentucky, where I went to dinner at a hotel and happened to look into the kitchen where a colored man was preparing food. At the sight of this black face, offset with those terrible white whites of the eye, I was unable to eat my dinner, and so I left the table and went to my room.*[23]

In their classical study of racial attitudes, Katz and Braly (1933) found that college students most frequently described Negroes as "superstitious, lazy, happy-go-lucky, ignorant, musical, very religious, stupid, physically dirty, naive, slovenly, and unreliable.[24] In a later study, Bayton (1941) utilized the same technique employed by Katz and Braly to compare the racial attitudes expressed by a group of black college students with the responses made by the white subjects in the Katz-Braly study. While there were some significant differences in the responses made by the two groups, both frequently described the Negro stereotypically.[25] *Bayton's study strikingly indicated how, in 1941, blacks accepted the perceptions of themselves which were held by whites.* As we have emphasized in this chapter, this is the crux of the problem involved in changing the black self-concept. Writes Bayton," . . . we see the Negro accepting these 'typings' [made by whites], including much of the one assigned to himself."[26] These are the parents of today's adolescents.

A number of studies indicate that adults' racial attitudes tend to persist through time. In 1950, Gilbert replicated the Katz-Braly study to determine the changes in students' racial attitudes after an eighteen-year period. He compared the responses of the subjects made in his study to those in the earlier study by Katz and Braly. Gilbert found that the same stereotypic adjectives which were used to describe the black man in 1932 were used in 1950. However, a smaller percentage of students described the black man stereotypically in 1950 than in 1933.[27] A study by Bogardus (1958) also suggests the persistency of racial attitudes. He compared the responses of a national sample of university students on his social distance scale in the years 1926, 1946, and 1956. On the basis of this comparison, Bogardus concluded that racial attitudes endure through time. While he noted some changes, in each of his studies the white race was more highly valued than others.[28]

More recent research also indicates that college students frequently describe blacks and other minority groups stereotypically. Ehrlich (1962) gave a sample of white college students a list of 71 statements and asked them to select those which described Negroes and Jews.[29] The students characterized the Jew "as a distinctive, cohesive economic elite, exclusive, and ethnocentric, and aggressive and exploitative in his relations with others." The black man was seen as "the classic primitive . . . irresponsible, lazy, and ignorant." Ehrlich also found that students who stereotype one racial or ethnic group tend to stereotype others. He concludes:

> *The content of Negro and Jewish stereotypes emerges distinctively and mutually exclusive. While the content is exclusive, the number of Negro stereotypes endorsed shows a moderately high correlation with the number of Jewish stereotypes endorsed. Generality is thus employed in the willingness to stereotype and the association between the number of stereotypes endorsed. Specificity is displayed in the consensual pattern of selective assignments of Negro and Jewish stereotypes.*[30]

The research discussed above utilized college and university students as subjects. Supposedly, college students and graduates make up the most sophisticated segment of the population. However, as this research indicates, college students frequently express negative attitudes toward black people and hold the pervasive stereotypes of them. *The typical college curriculum apparently does little to change students' racial attitudes and perceptions.* It is especially regretable that college students exemplify racist attitudes because they are the future leaders in society who will hold key jobs and make major decisions which will affect the lives of black and other poor people. College students are tomorrow's legislators, business executives, opinion molders, college professors, school administrators, and classroom teachers.

Research also indicates that classroom teachers typically have negative attitudes toward poor and black youth, and low expectations for them. Becker (1952) interviewed sixty teachers in an urban school system who felt that "slum" children "were difficult to teach, uncontrollable and violent in the sphere of discipline, and morally unacceptable on all scores from physical cleaniness to the spheres of sex and 'ambition' to get ahead."[31] Gottlieb (1964) found that white teachers dislike teaching urban black children much more than do black teachers.[32] When discussing problems in the school, the black teachers stressed the shortcomings of the physical plant while white teachers emphasized the faults of the children. White teachers frequently described black children as "talkative, lazy, fun-loving, high-strung, and rebellious." Black teachers most frequently described them as "fun-loving, happy, cooperative, energetic, and ambitious." Half of the white teachers in a study conducted by Clark (1964) felt that black students were innately inferior to whites and unable to learn in school. They believed that urban black schools should become custodial institutions and not remain educational institutions.[33]

The attitudes of white professional educators are manifested in the textbooks which they write and which children read in school. A comprehensive study of minority groups in textbooks completed in 1949 indicated that black Americans are frequently stereotyped in elementary and high school textbooks.[34] A study conducted by Marcus in 1961 indicated that few substantial changes had been made in social studies textbooks since the study reported in 1949. A number of more recent studies, including those by Stampp et al (1964), Golden (1964), Sloan (1966), the Michigan State Department of Education (1968), and the author (1969), have confirmed the earlier findings and indicate that no *significant* changes have occurred in textbooks in recent years.[35]

Many practices in the public schools exemplify the negative racial feelings of professional educators and contribute to the deflation of the black child's self-concept. The dilapidated buildings which house most black urban schools, racist counselors who believe that black children should not enter certain white-collar professions, and academic tracks are practices which tell black children that they are valued less than other children. However, perhaps no other factors are more significant in influencing the black child's devaluation of himself than negative teacher attitudes, low teacher expectations, and racist teaching materials. The seminal research by Rosenthal and Jacobson (1968) illuminates the cogent impact that teacher attitudes and expectations have on student performance. Students who are expected to learn tend to achieve in school; those who are not expected to learn become academic failures and dropouts.[36] *Teachers at all levels must believe that black children can achieve and communicate this belief to them if these students are to gain confidence in their ability and more positive self-concepts.* In addition to developing more positive attitudes and higher expectations for black students, teachers must change the curriculum so that it is more congruent with the needs of black students. Black children will be unable to value themselves highly as long as their people are continually degraded in the textbooks which they are forced to read and memorize.

In a society in which college students and educational experts exemplify racist attitudes and perceptions, we would expect the general populace to express similar feelings—perhaps in a more intense form. A recent national survey (1968) suggests that racist attitudes and Negro stereotypes are also pervasive in the larger society.[37] Thirty-three percent of the respondents felt that a separate country for black people is a good idea, 14 percent believed that white babies are born smarter than black babies, 43 percent indicated that blacks are not as civilized as whites, and 35 percent felt that policemen should shoot one or two rioters, after warning, as examples to other black people. This latter finding is especially alarming.

Other findings in this study indicate that *blacks often accept many of the stereotypes about themselves which are perpetuated by white society.* Ten percent of the black subjects and 11 percent of the white respondents felt that black people have looser morals than whites. The research on children's racial attitudes already discussed illuminates the extent to which black people accept the evaluations of themselves which are made by whites. Black children frequently reject colored dolls because they are painfully aware of the negative evaluations of black in American life and because they are rejected by white children.

STRATEGIES FOR CHANGE: THE REDUCTION OF RACIAL PREJUDICE

The urgent need to help both children and teachers modify and clarify their racial attitudes has been suggested. The discussion which follows explores, through research reviews, strategies which can be effectively used to change racial attitudes and thus contribute to the elevation of the black child's self-concept.

Changing Children's Racial Attitudes

There is a lamentable dearth of research on the modification of children's racial attitudes. Generalizations which can be derived from the research are extremely limited by the methodological flaws inherent in most of it. Most of the studies reported in the journals lack sufficient sampling techniques, controls, and clear descriptions of the experimental treatments. In only a few of them have attempts been made to control the teacher variable, which is a critical component in the modification of racial prejudice. While there are serious gaps in the research, *it does indicate that teaching materials and methods can affect youngsters' racial feelings and predispositions.*

Schlorff (1930) found that a civic course designed to "increase tolerance toward the Negro" augmented children's racial tolerance.[38] The contributions which the black man has made to American society were emphasized in the course. Randomly selecting the subjects, and some attempt to determine the effects of teacher attitudes would have enhanced the value of this study. Remmers found that teachers must make *specific* efforts to successfully change children's attitudes. He also found that attitude changes persist over time, although there is slight regression toward the pre-experimental attitudes immediately after the experimental treatment.[39] A group of students developed more negative attitudes toward the black Americans after seeing the movie, *Birth of A Nation* (1933), because the Negro is stereotyped in that film.[40] In an experiment by Campbell and Stover (1933), the group that used *pictures* experienced more positive gains in attitudes than groups without visual aids.[41]

One of the most frequently quoted studies on the affects of teaching materials on children's racial attitudes is the study reported by Trager and Yarrow (1952).[42] This experiment had a significant affect on children's racial feelings. All changes were in the expected directions. Children exposed to a democratic curriculum expressed more positive racial attitudes; those exposed to an ethnocentric curriculum developed more negative racial feelings. Trager and Yarrow summarize their study as follows:

> *The changes achieved in the experiment demonstrate that democratic attitudes and prejudiced attitudes can be taught to young children. The experiment contributes to an understanding of some of the important conditions which are conducive to learning attitudes. Furthermore, it is apparent that children learn prejudices not only from the larger environment but from the content of the curriculum and its values.* If democratic attitudes are to be learned they must be specifically taught and experienced.[43] *[Emphasis added]*

Research by Johnson and Litcher and Johnson confirms the Trager and Yarrow findings. Both studies support the postulate that teaching materials affect children's racial attitudes toward ethnic groups and themselves. Johnson (1966) studied the effects of a special program in black history on the racial attitudes and self-concepts of a group of black children. The course had a significant effect on the boys' attitudes. However, the effect on the girls' attitudes and self-perceptions was not significant.[44]

> *The Freedom School . . . seemed to have some effect on the boys in the areas of self-attitudes, equality of Negroes and whites, attitudes toward Negroes, and attitudes toward civil rights. That is, they became more confident in themselves, more convinced that Negroes and whites are equal, more positive toward Negroes, and more militant toward civil rights.*[45]

In a recently reported study, Litcher and Johnson (1969) investigated "the effects of multiethnic readers upon the racial attitudes of white elementary students."[46] On all posttest measures, the children who had studied multiethnic as opposed to all white readers expressed significantly more positive racial feelings toward blacks. The authors write convincingly, *"The evidence is quite clear. Through the use of a multiethnic reader, white children developed markedly more favorable attitudes toward Negroes.*"[47]

Research by Horowitz (1947) indicates that *contact* with minority group members does not necessarily result in more positive attitudes toward them.[48] He used a series of tests to ascertain the racial attitudes of several hundred white boys in New York City, urban Tennessee, and in the rural and urban areas of Georgia. Some of the New York children attended segregated white schools; others attended racially mixed schools. Horowitz concluded that there were no significant differences in the racial attitudes expressed by the children in these various settings. The Northern white boys evidenced as much prejudice toward blacks as the Southern white boys. Children residing in urban areas of the South were as prejudiced as those living in rural areas. Children attending racially mixed schools were as prejudiced as children attending all white segregated schools. The black boys in the sample indicated a preference for white over black. The author interpreted his findings to mean that "*. . . attitudes toward Negroes are chiefly determined not by contact with Negroes, but by contact with the prevalent attitude toward Negroes.*[49] [emphasis added] Placing black and white children in the same social setting will not necessarily make them feel better about each other. The social norms in the situation and the attitudes exemplified by the adult models are the significant variables.

Although subject to the limitations of the research, a number of generalizations can be derived from the literature on changing children's racial attitudes, some of which is reviewed above. *The research suggests rather conclusively that children's racial attitudes can be modified by experiences in the school designed specifically for that purpose.* However, specific instructional objectives must be deliberately formulated; incidental teaching of racial relations is apparently ineffective. Also, clearly defined teaching strategies must be structured to attain the objectives. Attitude changes induced by experimental intervention will persist through time, although there is a tendency for modified attitudes to revert back to the pre-experimental ones. However, the effects of the experimental treatment do not completely diminish. This finding suggests that intergroup education programs should not consist of one-shot treatments. Systematic experiences must be structured which are designed to reinforce and perpetuate the desired attitudes.

Visual materials such as pictures and movies greatly enhance the effectiveness of attempts to change racial attitudes.[50] An effective intervention program must utilize multiethnic teaching materials which present minority groups in a favorable and realistic fashion. Black history can be used to help children develop positive attitudes and self-perceptions. Contact with minority groups does not *in itself* significantly affect children's racial attitudes. The prevalent attitude toward different races and groups existent in the social situation is the significant determinant of children's racial feelings. The attitudes and predispositions of the classroom teacher are an important variable in a program designed to foster positive racial feelings and augment the black child's self-image.

Changing Adults Racial Attitudes

The bulk of research on ways to modify the attitude of adults is characterized by many of the same flaws which we have already noted regarding the research on changing children's racial feelings. This research is also frequently confusing and conflicting. Despite these shortcomings, we can derive at least *tentative* generalizations from a discussion of these studies which can guide action until more sound and less questionable research is available. We cannot defer action until conclusive research is done. The urgent need to change the perceptions and racial feelings of classroom teachers has been emphasized throughout this chapter.

For a number of years social scientists and educators regarded the study by Young (1928) as sufficient evidence that college courses in race relations were ineffective in changing students' racial feelings.[51] Young found that college undergraduates did not rank 24 racial and ethnic groups differently *after* a course on racial problems. However, it is possible that the students felt more positively toward *each* of the groups *after* the experiment. Young's instrument was not sensitive to this kind of attitude change. Droba (1932) studied the effects of a three month Negro sociology course on students' racial attitudes and concluded that only *marginal* positive gains in the students' attitudes resulted from the experience.[52] In a regular college course on the history of education, Bolton (1935) incorporated a special unit on Negro education and found no significant differences in the experimental group's pretest and posttest responses.[53] Brooks (1936), in a similar study, also reported only marginal gains from a special unit on racial problems in a regular course of study.[54] Smith (1939) and Ford (1941) attained more significant attitude changes with courses on the black American.[55]

While the attitude changes from lecture and discussion courses have been marginal, students who have participated in a number of different kinds of activities, such as committee work, workshops, and field work, have experienced significant positive gains in racial feelings. In an experiment conducted by Laird and Cumbee, (1952) college students studied the development of children's attitudes, participated in committees, discussions, and conducted informal surveys.[56] They also read literature on race relations, gave oral reports, made visitations, and listened to guest speakers. Significant positive changes in attitudes toward the blacks were observed in two of the four groups that participated in the experiment. The authors felt that the *committee work* was one of the strongest and most effective features of the experimental program.

A study by Smith (1947) lends considerable support to the postulate that adult racial attitudes can be modified by *contact and involvement* in minority group cultures.[57] * The experimental group toured Harlem, attended teas and dinners with black Harlem residents, and visited community institutions such as libraries and theaters. Lectures were also a part of the experimental program. The control group was not exposed to any of these experiences. Significant and consistently positive changes occurred in the experimental subjects' attitudes toward the black American. No significant changes were observed in the control subjects' racial feelings. Billings (1942) also studied the effects of a *field experience* and concluded that it had a positive effect on the students' racial attitudes and perceptions.[58] The experimental treatment consisted of weekly seminars in social problems, conferences which involved speakers who were specialists in various areas, and field participation.

Workshops have generally been commended for the contributions which they can make to an effective program to change racial

*While this finding seems to contradict the research by Horowitz reported earlier, it may be that the social norms in this experimental situation were significantly different from those in the social settings studied by Horowitz. It seems that racial and cultural contact can produce positive changes in racial feelings if it is coupled with pervasive norms which value all racial and ethnic groups.

attitudes. They can arouse the necessary concern about racism, enhance the dissemination of new ideas, equip teachers with effective teaching strategies, and change their personal attitudes.[59] Bogardus (1948) ascertained the effectiveness of a five-week intergroup education workshop on the participants' racial feelings.[60] The workshop consisted of lectures on racial problems of minority groups, research projects, and visits to community agencies. Bogardus concluded that the workshop had a significant positive effect on the participants' racial attitudes. The participants who scored in the middle range on the initial attitude scale were influenced most by the workshop experience. Those expressing extreme attitudes were the least affected.

Most of the attempts to change the racial feelings of adults have been limited to courses on minority groups either in regular classes or in workshop situations. A few efforts have been made to modify racial attitudes by providing experiences where individuals could have direct contact and personal experiences with minority group members. Some of these projects have involved the subjects in minority group community activities. *Psychotherapy* has not been extensively used to change racial attitudes. Haimowitz and Haimowitz (1950) reported an experiment in which positive changes in racial attitudes occurred as a result of participation in a "series of group therapy meetings."[61] The subjects were 24 counselors-in-training. A two-year followup study revealed that the attitude changes effected in the workshop were maintained. As in the Bogardus study, the individuals who scored in the middle range of the initial attitude scale were influenced greatest by the experimental treatment.

This research suggests that changing the racial attitudes of adults is a cumbersome task. To maximize the chances for successful intervention programs, experiences must be designed specifically to change attitudes. Programs with general or global objectives are not likely to be successful. Courses which consist primarily or exclusively of lecture presentations have little import. Diverse experiences, such as seminars, visitations, community involvement, committee work, guest speakers, movies, multimedia materials, and workshops, combined with factual lectures, are more effective than any single approach. Community involvement and contact (with the appropriate norms in the social setting) are the most cogent techniques. Psychotherapy is also promising. Individuals who express moderate rather than extreme attitudes are the most likely to change. This is encouraging since most individuals exemplify an average degree of prejudice.

THE CHALLENGE OF THE RACIAL CRISIS

The fact that our nation faces a racial crisis is now strongly apparent. We can see blatant evidences of this crisis from the elementary school to our college campuses. The power struggle over the public schools that occurred in the Ocean Hill-Brownsville district in New York City was in essence a racial battle. The black community was fighting to gain control over its public schools, while the city fathers were determined to keep the schools under their direct influence. The phrases "academic freedom" and "professional rights" were simply euphemisms designed to obscure the basic issues. The white community reacted strongly to the black community's bid for self-determination. Two traditional enemies, the Board of Education of New York and the New York Teachers Union, formed a coalition to resist a common enemy—the black community.

The racial crisis has also exploded on our college campuses. Black students are demanding that universities become more *relevant* to the black community. They want more black students admitted to universities, more black professors, courses related to the black experience, and on some campuses, separate living and eating facilities for black students. While a number of universities have granted concessions to black students, usually responding to force, some have bitterly resisted black demands. Throughout our society, white resistence to the black revolt is increasing, and the nation is becoming increasingly polarized. Whites are demanding "law and order" to stamp out the black revolt, and blacks are more determined than ever to secure their rights at any cost.

The schools must help to eliminate the urgent racial crisis which confronts us. The blood that ran in our black ghettos, the lives which have been lost, and the property which has been destroyed have not moved us much beyond the talking and committee stage. The government spent a small fortune on a report which told us what we have known for centuries, but did not want to admit, that we live in a racist society. Yet we have done little to eradicate racism since the National Advisory Commission on Civil Disorders Report was issued.

We have loosely and hurriedly put together a few courses on black history, hired black consultants to make brief visits to our schools, purchased a few "multiethnic" textbooks sprinkled with "selected" black heroes, and placed a few more black teachers and administrators in our schools. We have done most of these things after a violent racial outbreak or a racial assassination. We usually retreat back into *inaction* after the flames have cooled. In speaking before the Commission on Civil Disorders Kenneth Clark said:

> *I read that report . . . of the 1919 riot in Chicago, and it is as if I were reading the report of the investigating committee on the Harlem riot of '43, the report of the McCone Commission on the Watts riot.*
>
> *I must again in candor say to members of this Commission—it is a kind of Alice in Wonderland—with the same moving picture reshown over and over again, the same analysis, the same recommendations, and the same inaction.*[62]

We have talked and analyzed, but we have not *acted.* To help the black child augment his self-image and to save our society, we must first develop a commitment to deal with the root of the problem—*white racism.* Next, we must take serious steps to eliminate it. As educators, we must first eliminate the racism which pervades our own institutions. Colleges, and especially colleges of education, must implement bold and effective programs to change the racial attitudes of their students if we are to make any significant progress in enhancing the black child's self-concept in the near future. A classroom teacher who believes that his black pupils are physically and intellectually inferior will further deflate their already low self-concepts. It is imperative that we attempt to change teachers' racial attitudes in an era when racial polarization is so intense that policemen attempt to annihilate black militant groups which bitterly resist police harassment, and when a bigoted teacher can find "empirical" support for his belief in the intellectual inferiority of black people in the writings of an eminent and respected educational psychologist such as Arthur S. Jensen.

Jensen posits a conception of human intelligence which conflicts with the positions of most leading psychologists, although his work has been influential and has stirred up a controversy. He argues that human intelligence is basically genetically determined, that human beings differ in innate intellectual potential, and that whites are innately more intelligent than blacks.[63] Jensen's research is based on tenuous assumptions (he maintains that intelligence is what an IQ test measures). His argument is only an *hypothesis,* and should have never been presented to the general public (since it is only an hypothesis) in these racially troubled times. The hypothesis is **immoral, misleading,** and **irrelevant.** A moral hypothesis is that all children have untapped learning abilities, and we should search for the means to facilitate their learning. Jensen's argument only added fuel and ammunition to enflamed racial feelings.

We will be unable to change the attitudes of education students until we modify the perceptions and predispositions of their professors. Educationists and social scientists who have become *instant* experts on the black community exemplify the incalculable harm which can be done by scholars who possess racist attitudes. These "experts" and cultural deprivation theorists have created and perpetuated descriptions of black children which are little more than old stereotypes couched in new and sophisticated terminology—which makes them more insidious because they are legitimized and given a wide hearing in their classrooms, professional writings, and in the mass media. Jensen is more influential than George Wallace because he possesses academic credentials and does "rigorous research," although it is based on tenuous assumptions. These instant experts have described black children as "culturally deprived" and "anti-intellectual." Some have argued that they have "irreversible cognitive deficits" and are unable to perform "abstract and higher level cognitive tasks." As Clark has perceptively pointed out, the deprivation theories have been based primarily on speculation and conjecture and not sound empirical research. " . . .this concept [deprivation theory] has gained acceptance through intuition, general impressions, and repetition.[64] Clark writes further regarding white educationists who have become specialists on the learning problems of black children:

> *And when the North discovered its racism, it tended to provide justifications for it . . . in the academic community, it began to be clear in the 1960s that apparently sophisticated and compassionate theories used to explain slow Negro student performance might themselves be tainted with racist condescension. Some of the theories of "cultural deprivation," "the disadvantaged," and the like, popular in*

> *educational circles and in high govermental spheres until recently and in fact still prevalent, were backed for the most part by inconclusive and fragmentary research and much speculation. The eagerness with which such theories were greeted was itself a subtly racist symptom.*[65]

A teacher who accepts the deprivation theories and believes that his students have "irreversible cognitive deficits" will be motivated to do little to structure the kind of learning environment needed by his pupils, especially those who emanate from a culture which differs from his own and who he perceives as "lacking a culture." If teachers are to help black and other *culturally different* children to think more highly of themselves, they must understand that there are many ways of living and being, and that whether a cultural trait is functional or dysfunctional is determined by the social setting and situation. A middle-class white teacher who attempts to survive on the streets of Watts or in the midst of Harlem is as culturally disadvantaged as the black ghetto child in the white surburban school. Many of the cultural deprivation theorists have misled thousands of teachers into thinking that a child who possesses cultural traits which differ from theirs are without a culture or culturally deprived. American's oppressed ethnic groups have rich and diverse cultures. The phrase *culturally deprived* is one of the most unfortunate misnomers stipulated in the last decade. No educated person should have ever taken it seriously. The fact that it was invented and popularized by educational experts and theorists is a sad commentary on the quality of American educational leadership during the decade which recently ended.

A moratorium should be declared on *instant* white experts on the black community and on cultural deprivation and disadvantaged theorists before they contribute further to the deflation of the black child's self-image. Frank Riessman's book, *The Culturally Deprived Child,*[66] not only established him as an instant expert on black and poor children, but contributed as much to the perpetuation of racial and social class stereotypes as Ulrich B. Phillips' history of slavery in previous decades.[67] Reissman categorized poor children and described them as "deprived of culture," and "anti-intellectual":

> *The* **anti-intellectualism** *of the underprivileged individual is one of his most significant handicaps. It is expressed in his feeling that life is a much better teacher than books—that*

> *theory is impractical, "most big ideas that look good on paper won't work in practice," "talk is bull," intellectuals are "phony eggheads." This anti-intellectualism seems to be rooted in a number of the traits that characterize him; his physical style, alienation,* **antagonism to the school,** *defensiveness regarding his* **gullibility,** *and his generally pragmatic outlook.*[68] *[emphasis added]*

These unfortunate, categorical statements are merely speculations, unsupported by empirical data, yet they influenced many teachers who worked with poor and ethnic group children. Teachers who believe that their students are "anti-intellectual" cannot be expected to motivate them to achieve at high levels.

While white scholars have aggressively and persistently studied the effects of white racism as manifested in the black community, a regrettable paucity of work has been done in the last ten years on the nature and genesis of white racism. Perhaps so little empirical research has been done on the nature of white racism in recent years because most American scholars are white, and they have avoided a topic which may be personally uncomfortable, or it may be because more research funds were available for studying blacks than whites. It is not suggested here that a white scholar who has clarified and examined his racial attitudes cannot make a scholarly contribution to the research on the black community. However, since the characteristics of the black community and the black child have been extensively analyzed and described during the 1960s, current research should focus on the racial attitudes of whites and ways to change them.

The monumental studies on white racism by researchers such as Else Frenkel-Brunswik, Mary Ellen Goodman, Eugene L. Horowitz, and Gordon W. Allport were done prior to the 1960s.[69] The 1960s were characterized by a lack of research attention to white racism. However, the personality and pathologies of the black child and adult were thoroughly studied during this period, although much of this research was speculative, and based on racist assumptions. Since the pathologies in the black community merely mirror the problems of the larger white society, our greatest research need is to illuminate the origin and nature of white racism, which is the *root* cause of our major social problems and the black child's low self-concept.

Those white scholars who insist on studying the black

community despite current research needs, should at the least attempt to clarify their racial attitudes and formulate new assumptions, and at most become sensitive to the feelings and aspirations of black people. As professional educators, we will be unable to positively affect our students' racial feelings until we exemplify attitudes consistent with the American democratic ideology in both our teachings and writings. Teachers will not develop positive feelings toward their black children until professors become effective models. Many black children will continue to reject blackness as long as they have teachers who despise them and are coerced to read textbooks which they find infuriating. If serious steps are not taken *immediately* to eliminate racism in America we will not only fail to help black children develop positive self-concepts, but we may fail as a human society.

NOTES

1. Lines from the poem, "What Shall I Tell My Children Who Are Black," by Margaret Burroughs. Reprinted in Helen A. Archibald, ed., *Negro History and Culture: Selections For Use With Children* (Chicago: Community Renewal Society, undated), p. 4.
2. Quoted in Bradford Chambers, ed., *Chronicles of Black Protest* (New York: New American Library, 1958), p. 238.
3. National Urban League, *The Racial Gap 1955-1965: 1965-1975* (New York: National Urban League, 1967), Introduction.
4. Kenneth B. Clark, *Dark Ghetto* (New York: Harper & Row, 1965), pp. 63-67.
5. Robert L. Green, "Report Analysis: National Advisory Commission on Civil Disorders," *Harvard Educational Review* 38 (Fall 1968), p. 770.
6. Bruno Lasker, *Race Attitudes in Children* (New York: Henry Holt, 1929).
7. R. D. Minard, *Race Attitudes of Iowa Children* (University of Iowa: Studies in Character, vol. 4, no. 2, 1931).

8. Ruth E. Horowitz, "Racial Aspects of Self-Identification in Nursery School Children," *Journal of Psychology* 7-8 (1939), pp. 91–99.

9. Mary Ellen Goodman, "Evidence Concerning the Genesis of Interracial Attitudes," *American Anthropologist* 48 (1946), pp. 624–630.

10. Ibid., p. 626.

11. Kenneth B. Clark and Mamie P. Clark, "Racial Identification and Preference in Negro Children," in Theodore M. Newcomb and Eugene L. Hartley, eds., *Readings in Social Psychology* (New York: Henry Holt, 1947).

12. Helen G. Trager and Marian Radke Yarrow, *They Learn What They Live* (New York: Harper, 1952).

13. Ibid., p.136.

14. Ibid., p.150.

15. J. Kenneth Morland, "Racial Self-Identification: A Study of Nursery School Children," *The American Catholic Sociological Review* 24 (Fall 1963), pp. 231-242.

16. J. Kenneth Morland, "A Comparison of Race Awareness in Northern and Southern Children," *American Journal of Orthopsychiatry* 36 (January 1966), pp. 23-31.

17. Trager and Yarrow, p. 115.

18. Wilbur B. Brookover and Edsel L. Erickson, *Society, Schools And Learning* (Boston: Allyn and Bacon, 1969).

19. Helen H. Davidson and Gerhard Lang, "Children's Perceptions of Their Teachers' Feelings Toward Them Related to Self-Perception, School Achievement, and Behavior," *Journal of Experimental Education* 29 (December 1960), pp. 107-118. Reprinted in James A. Banks and William W. Joyce, eds., *Teaching Social Studies to Culturally Different Children* (Reading, Mass.: Addison-Wesley, 1971), pp. 113-127.

20. Dick Gregory (with Robert Lipsyte), *Nigger: An Autobiography* (New York: Pocket Books, 1964), pp. 31-32.

21. See the research by Howard S. Becker, David Gottlieb, and Kenneth B. Clark cited below. See also: David Gottlieb, "Goal Aspirations and Goal Fulfillments: Differences Between Deprived and Affluent American Adolescents," *American Journal of Orthopsychiatry* 29 (October 1964), pp. 934-941.
22. Emory S. Bogardus, "Social Distance and Its Origins," *Journal of Applied Sociology* 9 (1925), pp. 217-226.
23. Ibid., p. 222.
24. David Katz and Kenneth W. Braly, "Verbal Stereotypes and Racial Prejudice," in Newcomb and Hartley, pp. 204-210.
25. James A. Bayton, "The Racial Stereotypes of Negro College Students," *Journal of Abnormal and Social Psychology* 36 (1941), pp. 29-102.
26. Ibid., p. 101.
27. G. M. Gilbert, "Stereotype Persistence and Change Among College Students," *Journal of Abnormal and Social Psychology* 46 (1951), pp. 245-254.
28. Emory S. Bogardus, "Racial Distance Changes in the United States During the Past Thirty Years," *Sociology and Social Research* 43 (November-December 1958), pp. 171-176.
29. Howard J. Ehrlich, "Stereotyping and Negro-Jewish Stereotypes," *Social Forces* 4 (December 1962), pp. 171-176.
30. Ibid., p. 176.
31. Howard S. Becker, "Career Patterns of Public School Teachers," *Journal of Sociology* 57 (March 1962), pp. 470-477.
32. David Gottlieb, "Teaching and Students: The Views of Negro and White Teachers," *Sociology of Education* 27 (Summer 1964), pp. 345-353.
33. Kenneth B. Clark, "Clash of Cultures in the Classroom," in Meyer Weinberg, ed., *Learning Together* (Chicago: Intergrated Education Associates, 1964).
34. Committee on the Study of Teaching Materials in Intergroup Education, *Intergroup Relations in Teaching Materials* (Washington, D. C.: American Council on Education, 1949).

35. Lloyd Marcus, *The Treatment of Minorities in American History Textbooks* (New York: Anti-Defamation League, 1961); Kenneth M. Stampp et al., "The Negro in American History Textbooks," *Intergrated Education* 2 (October-November 1964), pp. 9-24; Loretta Golden, "The Treatment of Minority Groups in Primary Social Studies Textbooks," Ed.D. diss., Stanford University, 1964; Irving A. Sloan, *The Negro in Modern History Textbooks* (Chicago: American Federation of Teachers, AFL-CIO, 1966); Department of Public Instruction, *A Report on the Treatment of Minorities in American History Textbooks* (Lansing: Michigan Department of Education, 1968); James A. Banks, "A Content Analysis of the Black American in Textbooks," *Social Education* 33 (December 1969), pp. 954-957, ff., p. 963.
36. Robert Rosenthal and Lenore F. Jacobson, "Teacher Expectations for the Disadvantaged," *Scientific American* 218 (April 1968), p. 22.
37. *White and Negro Attitudes Towards Race Related Issues* (New York: Columbia Broadcasting System, 1968).
38. P. W. Schlorff, "An Experiment in the Measurement and Modification of Racial Attitudes of School Children," Ph.D. diss., New York University, 1930.
39. H. H. Remmers, "Propaganda in the Schools—Do the Effects Last?" *The Public Opinion Quarterly* 2 (April 1931), pp. 197-210.
40. Ruth C. Peterson and L. L. Thurstone, *Motion Pictures and The Social Attitudes of Children* (New York: Macmillan, 1933).
41. Don W. Campbell and G. F. Stover, "Teaching International-Mindedness in the Social Studies," *Journal of Educational Sociology* 7 (1933), pp. 244-248.
42. Trager and Yarrow, *They Learn What They Live.*
43. Ibid., p. 341.
44. David W. Johnson, "Freedom School Effectiveness: Changes in Attitudes of Negro Children," *The Journal of Applied Behavioral Science* 2 (1966), pp. 325-330.
45. Ibid., p. 129.
46. John H. Litcher and David W. Johnson, "Changes in Attitudes

Toward Negroes of White Elementary School Students After Use of Multiethnic Readers," *Journal of Educational Psychology* 60 (1969), pp. 148-152.

47. Ibid., p. 151.

48. Eugene L. Horowitz, "Development of Attitude Toward Negroes," in Newcomb and Hartley, pp. 507-517.

49. Ibid., p. 517.

50. Eunice Cooper and Helen Dinerman, "Analysis of the film 'Don't Be a Sucker': A Study of Communication," *Public Opinion Quarterly* 15 (Summer 1951), pp. 243-264.

51. Donald Young, "Some Effects of a Course in American Race Problems on the Race Prejudice of 450 Undergraduates at the University of Pennsylvania," *Journal of Abnormal and Social Psychology* 22 (1927-28), pp. 235-242.

52. D. D. Droba, "Education and Negro Attitudes," *Sociology and Social Research* 17 (1932), pp. 137-141.

53. Euri Belle Bolton, "Effect of Knowledge upon Attitude Towards the Negro," *Journal of Social Psychology* 6 (1935), pp. 68-88.

54. Lee M. Brooks, "Racial Distance as Affected by Education," *Sociology and Social Research* 21 (1936), pp. 128-133.

55. Mapheus Smith, "A Study of Change of Attitudes Toward the Negro," *Journal of Negro Education* 8 (1939), pp. 64-70; Robert N. Ford, "Scaling Experience by a Multiple-Response Technique: A Study of White-Negro Contacts," *American Sociological Review* 6 (1941), pp. 9-23.

56. Dorothy S. Laird and Carrol F. Cumbee, "An Experiment in Modifying Ethnic Attitudes of College Students," *The Journal of Educational Sociology* 25 (1952), pp. 401-410.

57. F. T. Smith, "An Experiment in Modifying Attitudes Toward the Negro," reported in Arnold M. Rose, *Studies in the Reduction of Prejudice* (Chicago: American Council on Race Relations, 1947), p. 9.

58. Elizabeth L. Billings, "The Influence of a Social Studies Experiment on Student Attitudes," *School and Society* 56 (1942), pp. 557-560.

59. Hilda Taba, "The Contributions of Workshops to Intercultural Education," *Harvard Educational Review* (March 1945), pp. 122-128.

60. Emory S. Bogardus, "The Intercultural Workshop and Racial Distance," *Sociology and Social Research* 32 (1948), pp. 798-802.

61. Morris L. Haimowitz and Natalie Reader Haimowitz, "Reducing Ethnic Hostility Through Psychotherapy," *The Journal of Social Psychology* 31 (1950), pp. 231-241.

62. Quoted in Green, p. 771.

63. Arthur R. Jensen, "How Much Can We Boost IQ and Scholastic Achievement," *Harvard Educational Review* 39 (Winter 1969), pp. 1-123. See also: "How Much Can We Boost IQ and Scholastic Achievement: A Discussion," *Harvard Educational Review* 39 (Spring 1969), pp. 273-356, for eight perceptive reactions to the Jensen hypothesis.

64. Kenneth B. Clark, "Fifteen Years of Deliberate Speed," *Saturday Review* (December 20, 1969), p. 60.

65. Ibid., p. 60.

66. Frank Riessman, *The Culturally Deprived Child* (New York: Harper & Row, 1962).

67. Ulrich B. Phillips, *American Negro Slavery* (New York: Appleton, 1918).

68. Riessman, p. 29.

69. See: T. W. Adorno, Else Frenkel-Brunswik, D. J. Levinson, and R. N. Sanford, *The Authoritarian Personality* (New York: Harper & Row, 1950); Mary Ellen Goodman, *Race Awareness in Young Children* (New York: Collier, 1952); and Gordon W. Allport, *The Nature of Prejudice* (Reading, Mass.: Addison-Wesley, 1954). Work by Eugene L. Horowitz is cited above.

All I wanted was to be a man among other men.
I wanted to come lithe into a world that was ours
and to help to build it together.
. . . I wanted to be a man, nothing but a man.[1]
Frantz Fanon

2

The Black Revolution and Education

Donald H. Smith

In the autumn that followed the great summer revolt of 1967, a new black student entered America's schools. His likes had never been seen before, and his coming was devastating. Public schools that had hardly known how to deal with his forerunners found him unfathomable.

His haircut and the ebony tiki around his neck were strangely African. Strange for a boy who in the recent past had rolled his eyes in embarrassment and looked away at the mention of Africa or blackness. During that fiery summer he had worn sandals and thought "black is beautiful," "I am my black brother's brother," and "it's so beautiful to be black." If he lived in Newark or Detroit or even one of the less scorched cities, he may have hurled bottled fire. Whether or not he participated in the riots, he is likely to have experienced a feeling of power and pride as he watched his peers lash out at society and its agents, the police and the fire fighters.

In their way, speaking to society with the only means they could discover, dispossessed black youth signaled their desperate determination to strike down the ghetto walls. And further, they signaled the assertion of their selfhood. Denied dignity and acceptance by the white society that had promised equality in exchange for assimilation, these youngsters were reversing the psychology of rejection and self-abnegation with the counter-psychology of beautiful blackness and self-love.

These young people, who burned and looted during the summer of 1967 and during the emotion-wrought period following the assassination of Dr. Martin Luther King, Jr., were attempting to survive. They fought for their lives in the only way they knew how. They enraged a nation that did not even know they existed. By some standards, the behavior of these young people is considered delinquent and dysfunctional. Others, however, label these disruptive responses as "normal" in an "abnormal" society, as healthy in a sick society. Kenneth Clark, for instance, has written the following:

> *The Negro delinquent, therefore, calls attention to the quiet pathology of the ghetto which he only indirectly reflects. In a very curious way the delinquent's behavior is healthy; for, at the least, it asserts that he still has sufficient strength to rebel and has not yet given in to defeat.*[2]

The schools of the nation that watched Watts, Newark, and Detroit burn without perceiving the true message of the flames have been equally imperceptive about the emergence of the new angry black students, the children of revolt, in their own corridors. Many urban schools have become battlegrounds for the open revolt of black pupils. Student uprisings have jolted schools and communities from their complacent attitudes about race relations. Schools throughout the nation have experienced open hostility in the form of violence between black and white students and between black students and school personnel. In all-black schools, students have declared war upon school personnel, mostly white, but including some black teachers. In some biracial schools, hostilities have erupted as a result of open conflict between individuals or groups of black and white students. In others, mere rumors of racial conflict have been sufficient to create the fact.

At first, the complaints of black students fell into two categories: denial of the opportunity to become homecoming queen or king and nonrepresentation on cheering squads. Two seemingly minor issues, they were the catalysts that sparked black student revolts.

But another issue has come to the surface: in all-black and biracial schools black students are demanding the inclusion of their own history and culture in textbooks and curriculum. But these issues—determination to participate in significant school activities and insistence upon the recognition of black culture and achievement—are part of a larger and more fundamental problem. *The revolt in the schools is a microcosm of the revolt of black people in American society.* And this revolt has changed from a nonviolent direction to one that includes violent conflict.

THE QUIET REVOLUTION

Black America has always been angry with white America (Kardiner and Ovesey, 1964; Lomax, 1963). Until quite recently, most blacks managed to sublimate their hostilities into channels that would not bring direct confrontation with the dynamite stick or the lynch rope. But an oppressive environment offers a limited number of responses: withdrawal, acquiescence, accommodation, and confrontation (Pettigrew, 1964).

Poor blacks who came into daily contact with whites and whose livelihood, to say nothing of life and limb, depended upon the continued approval of whites frequently accommodated themselves by grinning and scratching and mouthing the expected platitudes of survival. Still others built protective walls of apathy.

Blacks who were more secure financially, the teachers, professors, and physicians, particularly in the South, found their psychic survival in withdrawing from direct interaction with whites. Yet, when necessity dictated, they, too, swallowed pride and dignity and bowed to the vaunted intimidation of whites who are able to give jobs but also take away life (Frazier, 1957).

In spite of the protest activities of black militants such as Denmark Vesey and Nat Turner, who led slave insurrections, and in spite of hundreds of slave revolts, there had never been a sustained, united black confrontation with the social order prior to the 1950s.

The Montgomery bus boycott of 1955-56 provided the first opportunity for American blacks to confront oppression directly and massively. During the Montgomery confrontation, Dr. Martin Luther King, Jr., emerged as the nonviolent apostle whose philosophy and tactics dominated civil rights activities for over a decade.

Dr. King's philosophy of the social gospel was undergirded by the thinking of two men: Thoreau, who believed that men have a moral obligation to resist unjust laws and to accept the penalty for breaking such laws; and M. K. Gandhi, whose own practice of nonviolent "Soul Force" exacted India's freedom from the British.

The courageous victory won by Montgomery's blacks gave hope to their people all over the country, and Dr. King emerged as the most powerful black leader in American history. His work and that of his colleagues in the Southern Christian Leadership Conference forged important achievements during the 1960s, in terms of legislative accomplishments and positive shifts in the attitudes of white people (Hyman and Sheatsley, 1964). Following the demonstrations in Birmingham and the historic march on Washington in 1963, President Kennedy sent the strongest civil rights bill of all time to Congress and a year later President Lyndon Johnson guided the bill to passage.

One of the provisions of the Civil Rights Bill of 1964 attempted to abolish school segregation for the second time in a decade. This, coupled with multi-billion-dollar education bills, promised to improve substantially the educational opportunities of Afro-American children. Other provisions of the Civil Rights Acts of 1964 and 1965 had promised improved job opportunities and political power, but they failed to be effective; and, thus far, the education bills have failed to register any significant impact upon so-called "target area" schools. Measured by almost any standard—achievement scores, dropouts, window breakage, or attacks on teachers—life in urban schools grows worse, not better.

In spite of the inspirational leadership of Dr. King, the civil rights acts, the education bills, and the Economic Opportunity Act, the lives of poor people, like the lives of their children in the schools, have become more frustrating, more unbearable.[3]

Civils rights bills that are not enforced, poverty programs that fail to build self-determination and self-support, and the existence of institutional racism which relegates black citizens to the bottom of the heap—all combine to produce a strange mixture of anger and hatred, hope and despair, expectation and defeat.

THE MOVE TO VIOLENCE

The hope-despair syndrome, created by the great promises of legislation and the frustrating results, gave rise to the leadership of Malcolm X and the new black militants, typified by young revolutionaries Rap Brown and Stokely Carmichael. Curiously enough, Brown and Carmichael as young boys were under the tutelage of Dr. King and the SCLC. In Mississippi in 1964, Carmichael with much compassion schooled young blacks and whites in the philosophy and tactics of nonviolence. But that was before Carmichael, Brown, and the other young members of SNCC had endured the virtually unpunished murders of Medgar Evers, Viola Liuzzo, Jonathan Daniels, of Cheney, Schwerner, and Goodman, and before Carmichael was committed to Parchman Farm, the state penitentiary, for daring to encourage Mississippi blacks to register to vote. Such experiences can turn idealistic, compassionate young men into hardened realists. Such experiences can cause followers of nonviolence to take up the sword, rhetorically or sometimes in deed.

The black militants have given up on America's capacity to do what is right out of noble, humane motives. Instead, they believe that the nation must be forced to live up to its professed ideals. Corollary to forcing white America to deal justly with blacks, they believe the blacks must unite themselves into a self-protecting, self-determining, self-promoting group. Carmichael and Hamilton have called this process the closing of ranks:

> *By this we mean that group solidarity is necessary before a group can operate effectively from a bargaining position of strength in a pluralistic society.*[4]

Other militants move beyond the sound economic and political strategies suggested by Hamilton and Carmichael in *Black*

Power and into the realm of violent revolt, subscribing to the ideas of Frantz Fanon, the black psychiatrist from Martinique. Fanon believed that enslaved, colonized people must free themselves through violent revolution. As Fanon wrote in *The Wretched of the Earth:*

> *Violence is a cleansing force. It frees the native from his inferiority complex and from his despair and inaction; it makes him fearless and restores his self-respect.*[5]

Following the assassination of Dr. King, the poor-affluent, black-white schism was widened and the battle lines hardened. As Black Panther leader Eldridge Cleaver wrote:

> *The assassin's bullet not only killed Dr. King, it killed a period of history. It killed a hope, and it killed a dream. That white America could produce the assassin of Dr. Martin Luther King is looked upon by black people—and not just those identified as black militants—as a final repudiation by white America of any hope of reconciliation, of any hope of change by peaceful and nonviolent means.*[6]

White America has not been completely unconscionable and unconcerned about the plight of black America. Attempts have been made to redress some of the grievances, to bridge part of the gap, to repair some of the damage. Unfortunately, most of these efforts have been planned and administered by whites, and occasionally by blacks who have not represented the best interests of lower-class blacks. Never have commitments and funds been anywhere near the requirements; almost never have lower-income blacks, or concerned middle-class blacks, been in control. The contemporary black mood asserts that blacks are determined to take control.

Malcolm X, one-time Muslim minister and assassinated leader, is the martyred hero and inspirational symbol of the new black quest for self-determination. His philosophy of black nationalism is the guiding force of black militancy. As Malcolm X announced in 1964:

Our political philosophy will be Black Nationalism. Our economic philosophy will be Black Nationalism. Our cultural emphasis will be Black Nationalism.[7]

In explaining the substance of tripartite black nationalism, Malcolm X stated:

. . . the political philosophy is that which is designated to encourage our people, the black people, to gain complete control over the politics and the politicians of our community.

Our economic philosophy is that we should gain economic control over the economy of our own community, the businesses, and other things which create employment so that we can provide jobs for our own people instead of having to picket and boycott and beg someone else for a job. . . . our social [cultural] philosophy means that we feel that it is time to get together among our own kind and eliminate the evils that are destroying the moral fiber of our society, like drug addiction, drunkenness, adultery. . . . We believe that we should lift the level or the standard of our own society to a higher level wherein we will be satisfied and then not inclined toward pushing ourselves into other societies where we are not wanted.[8]

REVOLUTION IN THE SCHOOLS

Within the last decade, the civil rights movement has forced considerable attention upon the plight of Americans who hunger in a land of plenty. Puerto Ricans, Mexican-Americans, American Indians, and poor Southern whites have been thrust into the national spotlight along with black Americans. Since the early 1960s, verbiage in great profusion has described the characteristics of the dispossessed poor, their world views, and their pathology. Believing the old cliche that "education will cure the nation's ills," Congress has appropriated billions of dollars to save the children of poverty.

Tragically, the sum total of these efforts has left America's blacks and our other poverty-stricken children still outside of the educational mainstream and outside of the social mainstream. The panacea—compensatory education—has proved to be a colossal failure. Like so many efforts to help, or at least to placate the poor, compensatory education never had a chance. It is ludicrous that white teachers and administrators could believe themselves capable of devising special compensatory programs to do the job they were incapable of doing in the far more lengthy regular program (Smith, 1968).

However, failure has not discouraged whites from continuing to dominate and control the education of blacks; it has not prevented large cities from continuing to appoint white school superintendents to administer black and Spanish-American majorities.

Some whites are interested, perhaps deeply, in the educational needs of the nation's twenty-two million blacks. However, it is the belief of this writer, supported by the compelling evidence of nonperformance, that however well-meaning whites may be, they lack the social perception to penetrate the mass of white racism that permeates the American school but is almost imperceptible to them. Instead of addressing themselves to the real source of the failure to educate blacks, white educators have busied themselves with methods, techniques, and special curricula for "motoric" children.

The President's Commission on Civil Disorders has correctly identified white racism as the corrosive force which is rotting the American fabric. In the words of the commission:

> *What white Americans have never fully understood . . . but what the Negro can never forget . . . is that white society is deeply implicated in the ghetto. White institutions created it, white institutions maintain it, and white society condones it.*
>
> *Race prejudice has shaped our history decisively; it now threatens to affect our future. White racism is essentially responsible for the explosive mixture which has been accumulating in our cities. . . .*[9]

Nowhere is the effect of white supremacy more pervasive and more debilitating than in the American school. Whether it takes the form of textbooks which promulgate white supremacy by excluding the lives and accomplishments of blacks and other minorities, whether it takes the form of white teachers who have double standards of expectation, reward, and punishment, or whether it takes the form of self-hating black teachers who despise black children—white racism has poisoned the American school. White supremacy has left many black teachers and white teachers paralyzed in its wake; and it has been most deadly when they are unaware of their social sickness.

At last, black pupils have begun to discover that they must force the schools to serve their needs by taking the same kind of chaotic action within the schools that others are taking in the larger society.

The demands and actions of the black students confront the racism that has always been present in the schools and is finally being unmasked. The view of many administrators and teachers that student revolts are deleterious is refuted by others who see the efforts to attack and destroy racism as a positive force that will benefit both blacks and whites.

As Dr. Martin Luther King wrote in his last book, *Where Do We Go From Here: Chaos or Community?:*

> *The value in pulling racism out of its obscurity and stripping it of its rationalizations lies in the confidence that it can be changed. To live with the pretense that racism is a doctrine of a very few is to disarm us in fighting it frontally as scientifically unsound, morally repugnant and socially destructive. The prescription for the cure rests with the accurate diagnosis of the disease. A people who began a national life inspired by a vision of a society of brotherhood can redeem itself. But redemption can come only through a humble acknowledgement of guilt and an honest knowledge of self.*[10]

How incredible it is that young people have to threaten and sometimes bring about destruction to get the attention of society outside and inside the schools. Their goals are so amazingly simple, so undeniably just, that rational men must wonder about the wisdom, the morality, even the sanity of those who would deny their goals because the means by which they are communicated are discomforting.

WHAT DO THEY WANT?

First, what black pupils want and need are teachers who believe they can learn, who expect them to learn, and who teach them. Teachers whose naivete and cultural biases have conditioned them to believe that blacks, Indians, poor whites, or Spanish-speaking children are inferior can never teach them. It is extremely difficult for most teachers to understand how their own perceptions of the worth and ability of their students actually affect the emotional development and achievement of the children.

The research of Rosenthal and Jacobson (1967) suggests the critical relationship between teacher expectation and pupil achievement. In this study, involving an experiment with rats, graduate students received information that certain rats were "abnormal." As a result, the students failed to teach these rats to perform expected tasks, even though some of the rats so labeled were actually normal. Yet with groups of rats alleged "normal," the graduate students were successful in teaching the same tasks. The labels themselves became determinants of how the rats were perceived and of the subsequent behavior of the experimenters in attempting to teach their subjects.

Rosenthal and Jacobson's work with the children of Oak School in the South San Francisco Unified School District bore similar results. Based upon an achievement test and the random selection of certain children as "spurters" in achievement, Rosenthal and Jacobson were able to convince classroom teachers that the children so designated would undergo significant achievement spurts during the forthcoming year. The children labeled as spurters actually did excel because, as Rosenthal and Jacobson conclude:

> *. . . one person's expectation of another's behavior may serve as a self-fulfilling prophecy. When teachers expected certain children would show greater intellectual development, those children did show greater intellectual development.*[11]

A teacher need not be an avowed advocate of race or class supremacy to damage the emotional or intellectual growth of minority pupils. It is very possible for a well-intentioned teacher to succumb unwittingly to thinking that children who live in housing projects or slum tenements, who are supported by public assistance, whose skins are dark, or whose language is nonstandard are not able to learn. Such beliefs may cause teachers to despair at the hopelessness of it all or cause them to engage in the curious rationalization that because such pupils are unlikely to succeed, they would waste effort trying to teach them.

Let it be clearly understood that all children want to learn in school. Further, all mothers and fathers in the black ghetto want their children to receive a good education. Parents often are unable to communicate this desire, and frequently the horror of their lives forces them into acts which appear neglectful and unconcerned. The children themsleves enter school eager to learn, in love with their teachers, with policemen, with firemen, with everybody. Sometimes they, too, lack the signs by which their desire to learn might be communicated to teachers whose own culture has taught them how to detect readiness and willingness to learn.

Dr. Helen Redbird tells a classic story that underscores the absurdity of concepts such as "school-oriented children" and "readiness to learn." Dr. Redbird, professor at the Oregon College of Education, visited an elementary school to inquire about the progress of a little American Indian boy. She was told by the boy's teacher that he was doing poorly in class. The teacher explained that he appeared disinterested in learning, perhaps unwilling to learn. Dr. Redbird, who is an Indian, asked to remove the boy from the classroom for a few days to see if she couldn't get him "ready to learn." The teacher granted permission, and after a few days of "motivating" the student, Dr. Redbird returned him to his class.

About a week later Dr. Redbird inquired about the boy's progress. Not to her surprise, he was doing excellently and his teacher was amazed with the results. Unknown to the teacher, Dr. Redbird had taught the little boy two things: *To smile and to nod his head in response to his teacher.* When the boy had mastered these two acts of accommodation, his teacher was convinced of his willingness to learn! Neither smiling nor nodding was part of the little boy's culture, but until he was taught to behave in ways that signaled his desire to learn, he was abandoned. How many children are lost because they don't know or are unwilling to play the school game?

Second, black pupils need a curriculum that will release them from psychological captivity. The literature is replete with studies that reveal the psychological damage that slavery and post-slavery racism have imposed on black people. In spite of the growing Black Power, Black Pride movement, millions of blacks are trapped in the delusion of worthlessness so carefully engineered by an exploitative larger society.

When black children do not see themselves in their textbooks, when they are denied the chance to read of their people's accomplishments, when all about them they see only maids and porters, high rises and run-down tenements, when they perceive in the mass media only a replication of the black meniality and degradation that are their daily companions, they are trapped by self-doubt and self-rejection. Black children must have proof of their own worth. They must learn about their own worth as a derivative of the worth of their forebears. Black children must be taught to understand and appreciate their cultural heritage by teachers who understand and appreciate that heritage. Black children must know who they are and they must learn about the racial accomplishments of which they can be very proud.

Dr. Frantz Fanon, a victim of the "black is worthless" philosophy, came to realize that he had been seeking the wrong thing when he desired the approval of whites. He learned after much emotional turmoil that the only acceptance that has any meaning is self-acceptance:

> *I resolved, since it was impossible for me to get away from an inborn complex, to assert myself as a BLACK MAN. Since the other hesitated to recognize me, there remained only one solution: to make myself known.*[12]

As a concomitant to curriculum which is meaningful and inspiring, black pupils want to be taught and administered by models with whom they can identify and from whom they can derive feelings of pride and worth. Black pupils need to be taught and administered by educational personnel who are proud of black culture. They need to see their own people in control of their schools, not playing second or third man to "Mr. Charlie."

PROPOSALS FOR CHANGE IN THE SCHOOLS

Obviously, many readers of this chapter will be in agreement with it. However, many readers will find the ideas difficult to understand and even more difficult to accept. Some may dismiss the writer's interpretation of the contemporary black mood; others may agree with the interpretation but deny that white racism and oppression are responsible for black deprivation and subsequent anger and bitterness; still others may accept the interpretation, concur with racist causation, but believe that black nationalism is white racism in reverse.

For those who are convinced that the assertions in this chapter are invalid, the proposals that follow will have little meaning. For those who agree or are willing to suspend disbelief temporarily, these proposals may have positive impact.

The rationale for these proposals is based upon the following assumptions:

1. Racism is pervasive in all American institutions.

2. Racism in the American school is destroying black children and other minority youth, and little reversal appears likely without systematic counter-efforts.

3. American schools have served the interests and needs of the white middle and upper classes. Blacks, poor whites, Amerindians, and the Spanish-speaking minority have never been full members in the American school or the American way of life.

4. Black people have suffered deep psychological injury which has resulted in self-abnegation and group rejection.

5. Black students and their elders are justified in their anger, and they should be supported in their determination to gain equitable treatment.

6. The new wave of Black Consciousness and Black Pride is a positive psychological affirmation of the worth and dignity of Afro-Americans.

7. Ignored and denied approved channels to redress their grievances, black pupils have had no recourse but to be disruptive.

8. There are large numbers of educational personnel who are the unwitting agents of school systems that fail blacks and other minority pupils.

9. Given the opportunity, these educators would align themselves with other educators who possess the know-how and have the desire to educate minority pupils.

If the foregoing assumptions are valid, then the following proposals may be viewed as viable approaches for correcting the ills that deny black pupils their chance for equal participation in education.

1. Administrators, teachers, and other school personnel should undergo intensive sensitivity training to be able to engage in meaningful self-introspection. Hopefully, educational personnel would be helped to understand themselves and to discover how their own biases, stereotypes, and cultural limitations inhibit the emotional and intellectual growth of black pupils.

2. In addition to self-analysis, school personnel need to learn more about the historical, cultural, and educational characteristics of blacks and other poor. Courses such as the history and culture of Afro-Americans and other disadvantaged people should be requirements for undergraduate, graduate, and in-service training of educational personnel. Certainly these courses should have greater substance than the study of heroes and their deeds. They should come to grips with the fundamental issues of exploitation, oppression, and racism and various individual and group responses to those issues.

3. School personnel must be willing to abandon traditional curricular approaches that are questionable even for the white middle class and are totally inappropriate for nonwhites. By means of curriculum changes to include relevant political, economic, historical, cultural, and environmental experiences and materials, schools will succeed in the motivation of poor blacks, heretofore alleged as not possible. Nat Turner, Frederick Douglass, Ida B. Wells, Malcolm X, and Martin Luther King, Jr., should be the heroes of young black students. Their lives and deeds and those of other

outstanding Afro-Americans must provide the philosophy and psychology for black liberation. The politics and economics of social change must become the tools of physical liberation.

4. The schools must exercise deliberate and systematic efforts to provide equal educational and social opportunities *within* each school. Blacks who attend so-called desegregated schools must be afforded the opportunity and encouraged to participate in all academic and extracurricular activities. Even within an all-black school, if middle-class or "nice" conformist youngsters are most likely to be selected for activities, care must be taken to include children of the lower socioeconomic class.

5. Black youngsters need to be taught and administered by models of black manhood and black womanhood who have been released from white psychological captivity. In many central city areas, it is possible for black pupils to be taught by staffs that are predominantly black. The more difficult problem is for blacks to be administered and supervised by their own people. Big city school systems are not yet willing to bestow the control and high salaries of administrators upon many Afro-Americans.

6. In school systems where black pupils are a very small minority, it is probably unrealistic to expect a substantial number of black pupils to have black teachers. Obviously, for a number of years to come, some blacks will continue to be educated by white people, particularly when they live in integrated housing patterns. Such pupils will desperately need teachers who have been taught to understand and accept them as human beings and who are sensitive enough to the essence of black culture and the black experience to help black children appreciate themselves and their people.

7. Whether a school decides to integrate or to upgrade substantially the existing all-black schools, it must address itself to the issues expostulated in this chapter.

Considerable attention must be given to changing the attitudes of school personnel toward black pupils. This is true for black as well as white personnel, for a great many black teachers have been and remain the victims of vitiating white racism which causes them to demean and reject black children.

SUMMARY

After many years of being pushed out of schools, or dropping out of irrelevant, often hateful schools, black students are taking hold of their own destinies. They are in open rebellion against society and its agents, including the schools, that have kept the doors of opportunity closed and have treated them as a subhuman species. These young people are determined that they will be respected, that they will be taught, that they will have access to the same opportunities available to whites.

Whether the American schools recognize it or not, their black pupils are in revolt. They demand just treatment as well as relevant school experiences. Their anger and determination will be assuaged by nothing less than revolutionary responses—by nothing less than drastic changes in administrative and pedagogical attitudes and practices. Schools, like the greater society, cannot be maintained by positioning armed guards outside the doors. The schools and society must acknowledge their criminal neglect of black citizens, and they must take radical and forthright measures of correction. We must assume that the chaos in our school buildings and in our streets is a portent, not a final judgment. But we must understand that failure to act in massive and positive ways may be, in Dr. King's words, "mankind's last chance to choose between chaos and community."

NOTES

1. F. Fanon, *Black Skin, White Masks* (New York: Grove, 1967), pp. 112-113.

2. Kenneth B. Clark, *Dark Ghetto* (New York: Harper & Row, 1965), p. 88.

3. This conclusion has been reached by many, particularly the poor. See: the *Report of the National Advisory Commission on Civil Disorders*, 1968; Cloward, 1965.

4. S. Carmichael and C. V. Hamilton, *Black Power, The Politics of Liberation in America* (New York: Random House, 1967), p. 44.
5. F. Fanon, *The Wretched of the Earth.* Translated from the French by Constance Farrington (*Presence Africaine*, 1963; New York: Grove, 1966), p. 73.
6. From "Requiem for Nonviolence" in R. Scheer, ed., *Eldridge Cleaver* (New York: Random House, 1969).
7. Malcolm X, Announcement at Press Conference, March 12, 1964.
8. Malcolm X, Speech on Black Revolution, New York, April 8, 1964.
9. *Report of the National Advisory Commission on Civil Disorders*, pp. 2, 10.
10. M. L. King, Jr., *Where Do We Go From Here: Chaos or Community?* (New York: Harper & Row, 1967), p. 83.
11. R. Rosenthal and L. Jacobson, *"Self-Fulfilling Prophecies in the Classroom,"* 1967.
12. F. Fanon, *Black Skin, White Masks* (New York: Grove, 1967), p. 115.

REFERENCES

Cloward, Richard A., "The War on Poverty: Are the Poor Left Out?" *Nation* 201 (1965), pp. 55-60.

Frazier, E. Franklin, *Black Bourgeoisie* (Glencoe, Ill.: Free Press, 1957).

Hyman, Herbert H. and Sheatsley, Paul B., "Attitudes Toward Desegregation," *Scientific American* 211 (1964), pp. 14, 16-23.

Kardiner, Abram and Ovesey, Lionel, *Mark of Oppression* (Cleveland: World, 1964).

Lomax, Louis E, *When the Word is Given* (New York: New American Library, 1963).

Pettigrew, Thomas, *A Profile of the Negro American* (New York: Van Nostrand, 1964).

Report of the National Advisory Commission on Civil Disorders. U.S.

Riot Commission Report; also called the Kerner Report. Washington, D.C.: U.S. Government Printing Office, 1968.

Rosenthal, Robert and Jacobson, Lenore, "Self-Fulfilling Prophecies in the Classroom." Unpublished paper presented at the American Psychological Association, September 1967.

Smith, Donald H, "Changing Controls in Ghetto Schools," *Phi Delta Kappan* 49 (1968), pp. 451-452.

3

Black Youth and Motivation

Alvin Poussaint and Carolyn Atkinson

The civil rights movement of the 1960s in America spawned a new and vibrant generation of young black people. The degree of their commitment and determination came as a surprise for most white Americans, who, if they thought of blacks to any extent, considered them to be a rather docile, acquiescent people. As the events of the early 60s moved on with inexorable force, another, still younger generation of blacks stood on the periphery watching and waiting their turn. Their intense coming-of-age has brought still more surprise and puzzlement not only to white Americans, but to some Negroes. So busy have many been in "handling" or "coping with" the complex behavior of these young blacks, that relatively little has been done in terms of examining the basis for this behavior. This paper represents an attempt to pause in the on-going melee for an exploration of some of the factors of particular relevance to the motivation of Afro-American youth. Our analysis will focus on some of the "problem areas" so of necessity will not detail the many strengths and positive features of the black sociocultural environment. Of primary interest in our discussion will be the areas of internal motivation: the individual's self-concept; certain of his patterned needs; and the external motivators: the rewards offered by society for satisfactory performance in any of its institutional areas.

Of obvious importance to the functioning of any individual is his concept or vision of himself. And like it or not, this concept is inevitably a part of how others see him, how others tell him he should be seen. According to Mead,[1] Cooley,[2] and others,[3] the self arises through the individual's interaction with and reaction to other members of society: his peers, parents, teachers, and other institutional representatives. Through identification and as a necessary means of effective communication, the child learns to assume the roles and attitudes of others with whom he interacts. These assumed attitudes condition not only how he responds to others, but how he behaves towards himself. The collective attitudes of the others, the community or "generalized other," as Mead calls them, give the individual his unity of self. The individual's self is shaped, developed, and controlled by his anticipating and assuming the attitudes and definitions of others (the community) toward him. To the extent that the individual is a member of this community, its attitudes are his, its values are his, and its norms are his. His image of himself is structured in these terms. Each self, then, though having its unique characteristics of personality, is also an individual reflection of the social process.[4] This idea can be seen more succinctly illustrated in Cooley's suggestion of the self as a looking glass, a looking glass mirroring the three principal components of one's self-concept: "the imagination of our appearance to the other person; the imagination of his judgment of that appearance; and some sort of self-feeling, such as pride or mortification."[5]

For the black youth in white American society, the generalized other whose attitudes he assumes and the looking glass into which he gazes both reflect the same judgment: he is inferior because he is black. His self-image, developed in the lowest stratum of a color caste system, is shaped, defined, and evaluated by a generalized other which is racist or warped by racists. His self-concept naturally becomes a negatively esteemed one, nurtured through contact with such institutionalized symbols of caste inferiority as segregated schools, neighborhoods, and jobs, and more indirect negative indicators such as the reactions of his own family who have been socialized to believe that they are substandard human beings. Gradually becoming aware of the meaning of his black skin, the Negro child comes to see himself as an object of scorn and disparagement, unworthy of love and affection. The looking glass of white society reflects the supposed undesirability of the black youth's physical appearance: black skin and wooly hair, as opposed to the valued models of white skin and straight hair. In order to gain the esteem of the generalized other, it becomes clear to him that he must approximate this white appearance as closely as possible. He learns to despise himself and to reject those like himself. From the moment of this realization, his personality and style of interaction with his environment become molded and shaped in a warped, self-hating, and self-denigrating way. He learns that existence for him in this society demands a strict adherence to the limitations of his substandard state. He comes to understand that to challenge the definition the others have given of him will destroy him. It is impressed upon him that the incompetent, acquiescent, and irresponsible Negro survives in American society, while the competent, aggressive black is systematically suppressed. The looking glass of the black youth's self reflects a shattered and defeated image.

Several attempts have been made to determine how this shattered self-concept affects the black child's ability to function in society, his ability to achieve, to succeed or "make good," particularly in the area of education. Though the conclusions of these varied attempts have differed on occasion, there has been general agreement as to the reality of the black child's incomplete self-image.[6,7,8] One notable exception to this agreement, however, is the Coleman Report, a 1966 study by the United States Office of Education on the "Equality of Educational Opportunity."[9] This report maintains that the black child's self-concept has not been exceptionally damaged, and is in fact virtually no different from that of a white child. The report did, however, note that the white child is consistently able to achieve on a higher level than that of his black counterpart. The Coleman study, therefore, concluded that self-concept has little to do with an individual's ability to achieve.[10] Other studies tend to disagree with these findings.

Another variation on this theme of the black child's self-concept is seen in a 1968 report on "Academic Motivation and Equal Educational Opportunity" done by Irwin Katz. Katz found that black children tended to have exaggeratedly high aspirations, so high, in fact, that they were realistically impossible to live up to. As a result, these children were able to achieve very little:

> *Conceivably, their [low achieving Negro boys'] standards were so stringent and rigid as to be utterly dysfunctional. They seem to have internalized a most effective mechanism for self-discouragement. In a sense, they had been socialized to self-impose failure.*[11]

Katz presents evidence which indicates that the anticipation of failure or harsh judgment by adults produces anxiety in the child, and that in black children, this level of anxiety is highest in low achievers who have a high standard of self-evaluation.[12] Accordingly, a black child with an unrealistically elevated self-concept often tends to become so anxious concerning his possible failure to meet that self-concept that he does in fact fail consistently.

On the other hand, Deutsch's work has shown that Negro children had significantly more negative self-images than did white children.[13] He maintains that, among the influences converging on the black urban child,

> *. . . is his sensing that the larger society views him as inferior and expects inferior performance from him as evidenced by the general denial to him of realistic vertical mobility possibilities. Under these conditions, it is understandable that the Negro child would tend strongly to question his own competencies and in so questioning would be acting largely as others expect him to act, an example of what Merton has called the "self-fulfilling prophecy"—the very expectation itself is a cause of its fulfillment.*[14]

Similarly, Coombs and Davies offer the important proposition that:

> *In the context of the school world, a student who is defined as a "poor student" [by significant others and thereby by self] comes to conceive of himself as such and gears his behavior accordingly, that is, the social expecta-*

> *tion is realized. However, if he is led to believe by means of the social "looking glass" that he is capable and able to achieve well, he does. To maintain his status and self-esteem becomes the incentive for further effort which subsequently involves him more in the reward system of the school.*[15]

These views have been confirmed in such studies as that of Davidson and Greenberg.[16] In their examination of children from central Harlem, these authors found that the lower the level of self-esteem, the lower the level of achievement; while consequently, higher levels of self-appraisal and ego strength—feelings of self-competence—were associated with higher levels of achievement. For example, high achievers were more able to give their own ideas and to express basic needs, suggesting that a stronger self-concept is associated with a greater willingness to risk self-expression, an obvious prerequisite for achievement.

Certainly these various studies cannot be considered as ultimately nor unanimously conclusive. However, it is important to note that none of these reports has found any evidence of high achievement resulting from a low self-concept. Obviously the black child with such a low self-concept competes at a disadvantage with white youth in the struggle to achieve in this society.

The question then arises as to why black youth bother to involve themselves at all in this struggle. If their negative self-image handicaps them so greatly in achieving, why not simply abdicate and in fact adhere to society's definition of them as substandard? An attempted response to this question moves us into another area, that of patterned needs.

In the course of the socialization process, the individual acquires needs which motivate behavior and generate emotions. Three such needs concern us here: the need for achievement, the need for self-assertion or aggression, and the need for approval.

Among the attitudes of the generalized other which the individual in this society internalizes are the norms and values of the wider community, including, of course, the major tenets of the Protestant Ethic-American creed, i.e., with hard work and effort the individual can achieve success, and the individual's worth is defined by his ability to achieve that success. The individual who internalizes these values is motivated to act consistently with them, as his self-esteem is heightened or maintained through behaving in a manner approved by the community. Thus, the need for achievement develops in both white and black Americans. Consequently, the black youth's participation in the struggle for success is at least in part an attempt to satisfy his own needs.

This need to achieve may be very high as illustrated in the findings of Coleman[17] and Katz[18] who note the exceptionally high aspirations of Negro youth with regard to schooling and occupational choice. In addition, Katz[19] and Gordon[20] indicate that the aspirations and demands for academic achievement of the parents of these youth are also often exceptionally high. All of these sources agree, however, that the achievement of these youth is far from commensurate with either their own aspirations or those of their parents.[21] Thus, the problem does not seem to be, as some have suggested, one of insufficiently high levels of aspiration, but rather one of realizing these aspirations through productive behavior.[22] Gordon[23] and Katz[24] suggest that this discrepancy persists because the educational and occupational values and goals of white society have been internalized by black youth, but for one reason or another, the behavior patterns necessary for their successful attainment have not been similarly learned. Katz puts it succinctly:

> *Apparently the typical Negro mother tries to socialize her child for scholastic achievement by laying down verbal rules and regulations about classroom [behavior], coupled with punishment or detected transgressions. But she does not do enough to guide and encourage her child's efforts at verbal-symbolic mastery. Therefore, the child learns only to verbalize standards of academic interest and attainment. There standards then provide the cognitive basis for negative self-evaluations. . . . The low achieving Negro student learns to use expressions of interest and ambition as a verbal substitute for behaviors he is unable to enact. . . . By emphasizing the discrepancy between the real and ideal performance, anxiety is raised in actual achievement situations.*[25]

Thus, the black child's negative self-concept is further complicated by his internalization of white society's high-level goals, and the need to achieve them, without a true comprehension of how effectively to do so.

Further examination of the values of the Protestant Ethic leads to the conclusion that they imply that assertion of self and aggression is an expected and admired form of behavior. Through the socialization process, the individual internalizes those attitudes which reinforce his basic need to assert himself or express himself aggressively. Thus, random and possibly destructive aggression is channeled into a legitimate and rewarded avenue of achievement.[26]

What happens to the black child's need for aggression and self-assertion? What has been the nature of his socialization with respect to expressing aggression? Since slavery days and, to some extent, through the present, the Negro most rewarded by whites has been the Uncle Tom, the exemplar of the black man who was docile and nonassertive, who bowed and scraped for the white boss and denied his aggressive feelings for his oppressor. In order to retain the most menial of jobs and keep from starving, black people quickly learned that passivity was a necessary survival technique. To be an "uppity nigger" was considered by racists one of the gravest violations of racial etiquette. Vestiges of this attitude remain to the present day, certainly in the South, but also in the North: blacks who are too "outspoken" about racial injustices often lose their jobs or are not promoted to higher positions because they are considered "unreasonable" or "too sensitive." It is significant that the civil rights movement had to adopt passive-resistance and nonviolence in order to win acceptance by white America. Thus, the black child is socialized to the lesson taught by his parents, other blacks, and white society: don't be aggressive, don't be assertive. Such lessons do not, however, destroy the need for aggression and self-assertion.

One asserts oneself for self-expression, for achievement of one's goals, and for control of one's environment. Thus, an individual's success in satisfying his need for self-assertion is to some degree determined by his sense of control of his environment. Coleman found that of three attitudes measured, sense of control over environment showed the strongest relationship to achievement.[27] He further discovered that blacks have a much lower sense of control over their environment than do whites,[28] but that this sense of control increased as the proportion of whites with whom they went to school increased.[29] These findings indicate that for blacks, a realistic inability for meaningful self-assertion is a greater inhibitor of ability to achieve than is any other variable. These findings also suggest, however, that when blacks are interacting in a school situation which approximates the world in which they must cope, i.e., one with whites, their sense of control and achievement increases. Our emphasis here is not that black students' being in the presence of white students increases their sense of control and level of achievement, but that their being in a proximate real world suggests to them that they can cope in any situation, not just one in which they are interacting with others who, like themselves, have been defined as inferior.

Coleman's findings are supported by those of Davidson and Greenberg: high achievers were more able to exercise control and to cope more effectively with feelings of hostility and anxiety generated by the environment than were low achievers.[30] Deutsch points out that black male children for whom aggressive behavior has always been more threatening [compared with black girls] have lower levels of achievement on a number of variables than do black girls.[31] It is not surprising then that black people, objectively less able to control their environment than can whites, may react in abdicating control by deciding not to assert themselves. The reasons for this are clear. First, the anxiety that accompanies growth and change through self-assertion is avoided if a new failure is not risked and thus, a try is not made. Second, the steady state of failure through nonachievement rather than through unsuccessful trial is a pattern which many blacks have come to know and expect. They feel psychologically comfortable with the more familiar.

However, this effort by black people to deny their need for control and self-assertion inevitably takes its toll. Frustration of efforts to control the environment are likely to lead to anger, rage, and other expressions of aggression.[32] This aggression can be dealt with in a variety of ways. It can be suppressed, leading one to act on the basis of a substitute and opposing emotional attitude, i.e., compliance or docility. It can be channeled through legitimate activities—dancing, sports, or through an identification with the oppressor and a consequent striving to be like him. Aggression can also be turned inward and expressed in psychosomatic illness, drug addiction, or the attacking of those like oneself (other blacks) whom one hates as much as oneself. Or aggression can be directed toward those who generate the anger and rage—the oppressors, those whom the individual defines as thwarting this inclination to self-assertion. This final form of aggression can be either destructive or constructive: dropping out of school or becoming delinquent are examples of the former case, while participation in black social action movements is an example of the latter instance. This latter form of aggressive behavior amongst black people is increasing in extent. The old passivity is fading and being replaced by a drive to undo powerlessness, helplessness, and dependency under American racism. The process is a difficult one for those black people who manage to make the attempt. For their aggressive drive, so long suppressed by the ruling power structure, is exercised to the inevitable detriment of still another exigency: their need for approval.

With the development of the self and through the process of identification, the individual's need for approval develops and grows as does his need to avoid disapproval.[33] As we have stated earlier, the Protestant Ethic of American society approves behavior which follows the achievement motive and expresses the need for self-assertion. An individual's behavior in accordance with this ethic is often tied to a need for approval. On the other hand, for blacks in American society, the reverse is often the case, i.e., behavior which is neither achievement-oriented nor self-assertive is often approved by both blacks and whites (for different reasons), and thus, the need for approval may be met through behavior unrelated to either achievement or self-assertion.

Katz's study maintains that in lower-class black homes, children do not learn realistic (middle-class) standards of self-appraisal and therefore do not develop (as do middle-class children) the capacity for gaining "satisfaction through self-approval of successful performance."[34] Accordingly, Katz suggests that achievement should be motivated and rewarded by approval not from the home, but from fellow students and teachers.[35] The extent to which black children are responsive to approval for achievement in middle-class terms is, however, problematic. Some evidence suggests that lower-class black children are motivated to gain approval through physical characteristics and prowess rather than through intellectual achievement as are middle-class white and black children.[36] Further, needs for approval, not often met in black children through the established institutional channels, may be met by others outside of these legitimate institutional areas. For instance, delinquent subcultures support and encourage the behavior of their members. As a result, such members are not often sensitive to the informal sanctions imposed by nonmembers of this subsociety.[37] If an individual's needs are not met by others to whose sanctions he is expected to be responsive, he will be less likely to fear their sanctions for nonperformance and will seek to have his needs met by others to whose rewards of approval he will then be responsive.[38] Thus, for black youth no less than others, how the need for approval motivates behavior depends in large part upon how it is satisfied or rewarded.

The rewards which the institutions of this society offer to those whose behavior meets their approval or is "successful" consist of money, prestige, power, respect, acclamation, and love, with increasing amounts of each of these being extended for increasingly "successful" behavior. The individual is socialized to know that these will be his if he performs according to expectations. Hence, these rewards act as external motivators of behavior. Blacks have learned of the existence of these rewards. They have also learned, however, that behavior for which whites reap these rewards does not result in the same consequences for them. In the various institutional areas of society, blacks are often rewarded differentially from whites for the same behavior—if they are rewarded at all. How then can such a highly capricious system motivate their behavior?

That blacks orient some aspect of their behavior to society's reward system is evidenced by the fact that many studies have shown that lower-class blacks, as opposed to middle-class whites and blacks, have a utilitarian attitude toward education, viewing it primarily in terms of its market value.[39] The system provides no assurance, however, that once they obtain the proper education for a job, that they will in fact be allowed to get that job. This inability to trust society to confer rewards consistently no doubt makes it difficult for blacks to be socialized to behave in terms of anticipating future reward for present activity. Thus it is that Deutsch found that young black children are unwilling to persist in attempting to solve difficult problems. They respond to such situations with a "who cares" attitude.[40] Similarly, another study showed that when a tangible reward was offered for successful work on a test, the motivation of the deprived youngsters increased considerably.[41] In a New York program, young men who had been working primarily as clerks and porters were motivated to join a tutorial program for admission to a construction trade union apprenticeship program when they were promised that successful completion of the program (passing the union's examination) would definitely result in their being hired immediately at a salary often double what they were able to command previously.[42]

However, motivation to achieve certain rewards may have different consequences for behavior. As Merton explained, when the goals of society are internalized without a corresponding internalization of normative means for achieving these goals, what often results is the resort to illegitimate (deviant) means to achieve the socially valued goals.[43]

Just as a child unable to satisfy his need for approval through legitimate channels may turn to delinquent subcultures for support and encouragement, so too might such a child, unable to gain society's rewards by legitimate means, turn to illegitimate methods in order to attain them. Such forms of behavior as running numbers, pushing dope, and prostitution effectively serve to net the rewards of society, while circumventing the institutional channels for achievement of societal rewards. That Negro children early learn that such behavior is rewarded is suggested by Gordon's study in which young (9-13) central Harlem boys were asked if they knew people who had become rich, and if so, how they thought they had managed to do so. Of those who responded affirmatively, a majority felt that they had become rich through illegitimate means or luck.[44]

Consequently, for many black youth, external rewards are weak motivators of behavior, as they are discriminatorily and inconsistently given. The more immediate and direct the reward is, the stronger a motivator it is likely to be.

It would appear from this analysis that the standards and rewards of white American society simply do not work effectively to motivate productive behavior in young blacks. Clearly there is urgent need for a fundamental restructuring of the system. First, with respect to self-concept, all institutional segments of society must begin to function in a nonracist manner. To the extent that the self is shaped with reference to a generalized other, to that extent will the black child's image be impaired as long as America remains racist. The growth of black consciousness and pride have had salutary consequences for the black's self-image. But this alone is not sufficient. The operation of self-image as a motivator for behavior is like a self-fulfilling prophecy: blacks are continuously told and some believe that they are inferior and will fail. Therefore, they fail. For the black child to be motivated to achieve in school, the school must negate everything that the society affirms: it must tell the child that he can succeed—and he will.[45]

The relationship between self-concept and achievement is not clear-cut, but it appears to be a weaker motivator of behavior than the motive to self-assertion and aggression. More attention should be given to examining this dimension of personality as a motivator of the black youth's behavior than to continuing inquiries into his self-image. It has been noted that the black youth's sense of control of his environment increases as the proportion of whites in his school increases. It is imperative to keep in mind, however, that participation in all-black or predominantly black structures need not be self-destructive if the black youth chooses rather than is forced to participate in them. For if he chooses, he is asserting control over his environment. Those structural changes being made in American society in the direction of blacks having the opportunity to be more aggressively in control of their environment must be continued and expanded. The plans to decentralize New York City schools, to develop black business, and to organize and channel black political power are significant steps in this direction.

Most of the data indicate that black youth and their parents have high educational and occupational aspirations, which are not carried through to achievement levels. The reward systems of American society are often irrelevant to the lives and aspirations of most black youth. Approval is rewarded primarily for forms of behavior in which the black youth has managed to achieve little proficiency, making him less likely to make the effort. Something is obviously wrong with any school system which permits so much young potential to be wasted simply because it cannot be developed within the confines of traditional methods. New frameworks must be developed which will enable the educational aspirations of black youth to correspond to their interests and proficiencies. With the establishment of a pattern of consistent reward, there is every possibility that intellectual endeavors would have immediate relevance to their lives.

Certainly these suggested changes are sweeping, but so too have been the dangerous effects of the maintenance of the old systems. The time for being surprised at the behavior of black youth has passed. The time for lengthy, nonproductive attempts at understanding them has too, in its turn, come to an end. The time remaining must be effectively used in action to bring about these and similar changes. America cannot afford to wait for the next generation.

NOTES

1. George H. Mead, *Mind, Self, and Society* (Chicago: University of Chicago, 1934), Part III.
2. Charles H. Cooley, *Human Nature and the Social Order* (Glencoe, Ill.: Free Press, 1956), passim.
3. *Sociological Quarterly* 7, no. 3 (Summer 1966) (entire issue).
4. Mead, op. cit.
5. Cooley, op. cit., p. 184.
6. Alvin F. Poussaint, "The Dynamics of Racial Conflict," *Lowell Lecture Series,* sponsored by Tufts-New England Medical Center, April 16, 1968.
7. Joan Gordon, *The Poor of Harlem: Social Functioning in the Underclass,* Washington, D.C.: Report to the Welfare Administration, July 31, 1965, pp. 115 and 161; and Irwin Katz, "Academic Motivation and Equal Educational Opportunity," *Harvard Educational Review* 38 (Winter 1968), pp. 56-65.
8. While we recognize the limitations of many measures of self-concept and that self-concept is often defined by how it is measured, an exploration of these considerations within the scope of this paper is clearly impossible. Therefore, for the purpose of our presentation here, we are taking the measures of self-concept at face value.

9. James S. Coleman and others, *Equality of Educational Opportunity,* Washington, D.C.: U.S. Office of Education, G. P. O., 1966, p. 281.
10. Ibid., p. 320.
11. Katz, op. cit., p. 60.
12. Ibid., pp. 61-62.
13. Martin Deutsch, "Minority Groups and Class Status as Related to Social and Personality Factors in Scholastic Achievement," in Martin Deutsch and Associates, *The Disadvantaged Child* (New York: Basic Books, 1967), p. 106.
14. Ibid., p. 107.
15. R. H. Coombs and V. Davies, "Self-Conception and the Relationship between High School and College Scholastic Achievement," *Sociology and Social Research* 50 (July 1966), pp. 468-469.
16. Helen H. Davidson and Judith W. Greenberg, *Traits of School Achievers from a Deprived Background* (New York: City College of the City University of New York, May 1967), pp. 133, 134.
17. Coleman, op. cit., pp. 278-280.
18. Katz, op. cit., p. 64.
19. Ibid., pp. 63-65.
20. Gordon, op. cit., p. 115.
21. Coleman, op. cit., p. 281; Katz, op. cit., p. 63; Gordon, op. cit., pp. 155, 160-161.
22. David P. Ausubel and Pearl Ausubel, "Ego Development Among Segregated Negro Children," in A. Harry Passow, ed., *Education in Depressed Areas* (New York: Teachers College, 1963), p. 135.
23. Gordon, op. cit., pp. 115, 161.
24. Katz, op. cit., p. 63.
25. Ibid., p. 64.
26. Davidson and Greenberg, op. cit., p. 58.
27. Coleman, op. cit., p. 319.

28. Ibid., p. 289.

29. Ibid., pp. 323-324.

30. Davidson and Greenberg, op. cit., p. 54.

31. Deutsch, op. cit., p. 108.

32. Alvin F. Poussaint, "A Negro Psychiatrist Explains the Negro Psyche," *New York Times Magazine,* August 20, 1967.

33. Davidson and Greenberg, op. cit., p. 61.

34. Katz, op. cit., p. 57.

35. Ibid.

36. Edmund W. Gordon and Doxey A. Wilkerson, *Compensatory Education for the Disadvantaged* (New York: C. E. E. B., 1966), p. 18.

37. Claude Brown, *Manchild in the Promised Land* (New York: Macmillan, 1965), passim.

38. Talcott Parsons, *The Social System* (Glencoe, Ill.: Free Press, 1951), chap. 7.

39. Gordon and Wilkerson, op. cit., p. 18.

40. Deutsch, op. cit., p. 102.

41. Elizabeth Douvan, "Social Status and Success Striving," cited in Frank Reissman, *The Culturally Deprived Child* (New York: Harper & Row, 1962), p. 53.

42. Personal communication (C.A.).

43. Robert K. Merton, *Social Theory and Social Structure* (Glencoe, Ill.: Free Press, 1957), chap. 4.

44. Gordon, op. cit., p. 164.

45. Kenneth B. Clark, *Dark Ghetto* (New York: Harper & Row, 1965), pp. 139-148.

4

Political Socialization of Blacks: Implications for Self and Society

Bradbury Seasholes

Political effectiveness is a major key to social and economic betterment, and improvement in the social and economic spheres in turn further enhances that effectiveness. The union movement demonstrated this to the American laboring class. Political action, in conjunction with strictly economic action, helped win for union members better incomes and working conditions and also greater status and respect in society. And the greater income and respect in turn simply added to the unions' increase in political effectiveness. The use of *politics* as a major tool to stimulate this increase is instructive. It was the CIO, deeply committed to political action as the AFL was not, which was primarily responsible for labor's surge in the 1930s and 1940s.

Political effectiveness is also a key to personal stability. It helps to establish and maintain strong positions of income and respect against what would be, without the element of politics, the shifting winds of market and status. And political effectiveness reinforces an individual's general sense of personal adequacy in the society at large.

Because of these salutary effects on the objective and subjective lives of people, it is of greatest concern that in American society those who perhaps most need these consequences, blacks, have tended to be less involved and less effective as citizens than whites. Over the decades low involvement and low effectiveness in government have severely weakened a critical link, politics, between what blacks want in all major areas of life and the fulfillment of those aspirations. Now, in the 1970s, both rational and irrational activism have emerged as a powerful and drastically different mode of political behavior for significant numbers of blacks, creating a stark contrast and tension within black America as well as between the two races.

Neither low involvement nor militancy are characteristics of blacks which suddenly appear at adulthood. They are the behavioral consequences of states of mind that begin developing in early childhood. Just as we can speak of a child's gradual adaptation to society as "socialization," so we can speak of his increasing familiarity with his political environment as his "political socialization." Somewhere along the line, political socialization takes at least marginally different paths for blacks and for whites. In an attempt to understand how and why, it may be helpful to consider analytically what is encompassed for any individual, black or white, in the process of becoming socialized politically.

COMPONENTS OF POLITICAL SOCIALIZATION

Becoming politically socialized can be thought of as comprised of at least five component processes:

1. Learning to be *partisan.* Somewhere in one's life span, positions are taken, are to one extent or another made consistent with one another on issues, and are related to candidates running for office.
2. Learning to *participate* in politics. Beyond learning to take sides, a person develops a proclivity to participate in politics in certain ways (voting, wearing buttons, giving money, demonstrating in the streets, and so forth) and with a certain frequency or intensity.

3. Learning to be *optimistic* about politics. This refers to one's basic sense of effectiveness in politics. (It does not refer to whether or not one actually is effective.)
4. Learning political *information.* A large part of political socialization, of course, is accumulating facts about the structure of government, the avenues of activity open, the positions candidates and parties take on issues, and so forth.
5. Learning to participate with *skill.* Given a potential effectiveness in politics, how skillfully does a person use his resources to approach the full potential?

POLITICAL SOCIALIZATION IN THE SCHOOL YEARS

Partisanship, participation, optimism, information, and skills are learned (if at all) over a large number of years, and for most of us the process starts quite young. The literature available on political socialization concentrates on school-age children and adolescents, leaving an impression that the process is essentially complete by the early twenties. Although this is not true, certainly a large part of political attitudes and behavior patterns is developed prior to adulthood, that is, prior to the time when individuals are conventionally permitted to be full participants in government. In discussing what is currently known about political socialization of blacks, it will be necessary in large part to unite on a speculative basis two sets of information, one dealing with black adults and one with white political socialization.[1]

Regarding the learning of *partisanship,* we would expect that as a child proceeds toward adulthood and as his information about the world in general, and more particularly about the world of politics, increases, a sharpening of political opinion and ideology would occur. This sharpening would be noticed first through decreasing "no opinion" among school students and then through increasing polarization of political feelings within the growing group that does hold opinions. This hypothesis is pretty well substantiated. Herbert Hyman cites definite evidence that class differentials on political opinions do, in fact, increase over time as a student proceeds through his school years.[2] The basic reason is increased awareness of class positions and class interest, which leads to the finer definitions of interest translated into political stands.

Over a period of time during the school years, an increased differentiation develops in the amount of political *information* learned. That is, certain subgroups of students—the wealthy, for example—learn political facts at a greater rate than other subgroups. A Philadelphia study indicates that very young blacks are slower (in the aggregate) in learning what the name of their country is, and in recognizing such national symbols as the American flag and the Statue of Liberty, than white children of roughly comparable social class.[3]

What can we say about *participation?* One might expect that a growing awareness of one's specific class or other subgroup interest in the activities of a society would be coupled with a growing awareness of how the political process can have some effect on those interests. Consequently, the hypothesis would state that as one proceeds through the school years, greater interest in taking part in politics would accompany greater awareness of one's stake in society. But we run into some counterevidence, or at least counterspeculation, which points to a peaking of interest in participation and optimism about the effectiveness of participating somewhere before the completion of high school and the intervention thereafter of declining *optimism* about political efficacy.[4] Encroaching cynicism in turn is alleged to dampen interest in participation itself.

Among the very young it is difficult to distinguish fact-learning from affective development of optimism about governors. David Easton and Fred Greenstein have both published discussions of relatively young children's political information and feelings about political participants. With Robert Hess, Easton reports, for example, that children in grades two through eight tend to have very favorable images of the President and to have no knowledge or favorable feelings about other governmental officeholders. As children proceed from grades two through eight, their collective image of the President proceeds from being extremely like their feelings and impressions of their fathers to a position where father and President are quite distinct creatures. The increased differentiation comes at the expense of the father. For example, the child remains impressed at how much a President knows but becomes increasingly unimpressed with what his father knows.[5] Greenstein footnotes the Easton study by pointing out that in his quite similar work in New Haven, Connecticut, the mayor appeared quite prominent to young people and also had quite a favorable image. This he attributes to a particular situation, namely that the then mayor, Richard Lee, annually visited virtually every classroom in the city![6]

Greenberg's Philadelphia research documents the incidence and growth of clear black-white differences "at the margin" in feelings about the United States.[7] Although both races exhibit overwhelming loyalty, blacks are consistently less enthusiastic than whites.

All three of these studies leave some fascinating questions unanswered. Are black children of this age level more likely to have more information about and be more proud of black officeholders above and beyond their admiration for the President or possibly instead of admiration for the President? It is unlikely that many young school children know the names of, or care about, Congressmen, city councillors, or mayors; but are black children more likely to be aware of black officeholders at this level than white children are of whites? The extent to which black youngsters single out black athletes for special attention and admiration is well known. Is the same kind of excitement generated in them by blacks active in politics and government?

Sense of efficacy, frustration, and self-image

Central to the development of all the components constituting political socialization is a feeling that participating, taking stands, accumulating information, and sharpening skills are all worth the effort. A person is not likely to spend much time, energy, or money in politics if he is convinced that no results follow ("you can't fight city hall"), or that in some sense no results *should* follow ("I haven't got anything worth saying or being considered"). The first of these attitudes could be labeled conspiratorial and the second identified as the political segment of negative self-image. Traditionally, both have been considered under the general heading of a low sense of political efficacy.[8] But for the moment, it may be useful to consider the two separately.

"You can't fight City Hall"

That blacks have experienced severe frustration in political life is hard to dispute. In the South, city hall and especially the county courthouse have proven impenetrable fortresses to the brave few blacks who have in many instances literally risked their lives in trying to register their political preferences. As a country increasingly absorbed in our own urbanization and in the problems of blacks in the large cities, we tend to overlook how many blacks still live on farms in the South and how much their political thought and activity (or lack of it) contributes to the parameters of black politics nationally.[9] In the North, the actual frustration experienced by blacks is much more a function of social and economic disadvantage.

Presumably the experience of political frustration is one of the psychological mechanisms that ultimately account for low involvement of blacks in politics. But the relationship is not such a simple one. In theory, sense of frustration is designated as a state of mind resulting variously in destructive aggression, constructive aggression, and withdrawal. In short, frustration is unsatisfactory as a predictive or analytical device when taken as a total concept. It can account for apathy and militancy simultaneously.[10]

Some differentiation among kinds of frustration helps to explain the direction in which resulting behavior is likely to go. Three typical varieties of frustration known to blacks can be characterized as follows:

> 1. Achievement drive smothered by paternalism. Black orientation toward political achievement, or just plain perseverance, has in some instances been consciously or unconsciously undermined through paternalistic guidance and assistance by whites. Examples are not hard to find. The behavior of the big city machine is one (and suggests that blacks are not the only ethnic minority that has been "pastoralized" in this fashion). The attitude of "enlightened Southern white leadership" in systematically short-circuiting the growth of black leadership through token anticipation of black demands is another. And the way some whites once behaved in racially mixed integration organizations is a third example.
>
> 2. No help, and no help intended. The most typical Southern situation is severe frustration without accompanying encouragement toward self-help. This is frustration designed to perpetuate demoralization.

3. Frustration turned to constructive aggressiveness. On rare occasions, political leadership among blacks has been fostered through encouragement without paternalism in a general context of political frustration.

In the first instance, paternalism, the ultimate objective is to maintain black passivity, even if substantive concessions in public policy have to be made. In fact, concessions are used as a means of obtaining "behavior control." In the second instance, whether blacks react by becoming passive or aggressive is almost beside the point. The ultimate objective is to maintain a substantive policy status quo in which blacks are at a disadvantage. Inability of blacks to change their status by means of switching from aggressive to passive behavior constitutes "fate control" by whites. The two situations are illustrated in Figure 1.

Plus and minus signs represent states of mind of the dominant white population. Thus, in the behavior control situation whites feel good when they pass pro-black legislation *and* blacks remain or become passive, but feel bad if blacks respond to the legislation by

Whites:

Blacks are:	pass pro-black	reject legislation
passive	+	–
aggressive	–	+

BEHAVIOR CONTROL

Whites:

Blacks are:	pass pro-black	reject legislation
passive	+	–
aggressive	+	–

. . . or, at whites' discretion:

Blacks are:	pass pro-black	reject legislation
passive	–	+
aggressive	–	+

FATE CONTROL

Figure 1. Types of white control over blacks.

being aggressive. Whites will react to that latter condition by switching to their other plus sign; that is, they will answer black aggression by withholding (further) pro-black legislation.

In the fate control situation, the policy outcome is the same, regardless of what blacks do. This condition can arise when whites are simply oblivious to black desires and behavior, or, more malignly, as a conscious technique of creating anxiety and demoralization through unpredictable arbitrariness.

Although the third form of frustration is encountered all too rarely, it is a hope upon which to build. Fortunately, the condition need not presuppose conscious white policy of seeking to help Negroes in this matter. The condition may exist without the whites knowing it or wanting it. American Negro leaders today find themselves trying desperately to avoid falling prey to paternalism while in the process of escaping the total frustration of "no help, no help intended." For this reason, they now almost universally reject well-meant offers of white assistance. Of the alternatives open, they seem to prefer that of controlled frustration.

That controlled frustration can be effective in producing demonstrable increases in black interest, participation, and accomplishment is perhaps better illustrated in the sports than in the political realm. The disproportionate success of black youth in sports must be attributed to greater achievement drive in this area rather than to unusual physical characteristics of the race. (This is certainly the lesson of previous waves of stellar athletes drawn from other oppressed minorities earlier in the century.) There has been no particular white encouragement of black athletes—particularly ten to fifteen years ago—but once accomplishment has been made, there has been white admiration and approval.

Why has there not been so spectacular a rise in black involvement and success in political life? Except in the South, the frustration level has been about the same as in sports, that is, grudging permission by whites to take part, but no enthusiastic encouragement. On the other end, however, the parallelism sags. Political accomplishment by blacks is not generally followed by white admiration and approval. The athlete's reward—white acclaim and often increased social and economic well-being—is not lost on the black masses. The black politician or political activist is not generally so fortunate. To the black masses who might be tempted to adopt him as an idol, the rewards from political activity and high achievement are far less obvious than with their black sports idols. To be perfectly candid, they are not only less obvious; they may be less substantial in reality. For blacks eager to improve their images in others' eyes (and their own), political activity is not a clear-cut means to the end. The significance of "city hall" needs to be established if we are to expect anyone to bother with "fighting city hall"—or fighting *for* city hall. Recent accession of blacks to the office of mayor in major cities, heralding an era in which black control over the governance of central cities in America will become commonplace, is beginning to provide the test of the proposition that political visibility can affect political interest and attitude.

"I haven't got anything worth saying"

In the end, the most serious consequence of black frustration in politics is the possibly deleterious effect on blacks' self-evaluation. The black who sees politics as a conspiracy against him may or may not have a low political self-image. The black who traces his political insignificance to his own shortcomings does: "They don't care *because I am worthless.*" This quotation incorporates a big leap psychologically for the individual as well as theoretically. But empirically, the link between pessimism about government and a low self-image seems to be a strong one. This link can be seen in the lower right corner of Figure 2. The other cells describe remaining combinations of negative and positive self-images and negative and positive feelings toward government. In the diagram the likely—but not inevitable—behavioral consequence of each combination is given in parentheses.

	Responsive Government	Unresponsive Government
Positive Self-image	++ "I am worth a lot, and the government is responsive (so I *will* participate in politics)."	+− "I am worth a lot, but the government is not responsive (so I won't participate)."
Negative Self-image	−+ "The government is responsive, but I am worthless (so I won't participate)."	−− "I am worthless and the government is not responsive (so I won't participate)."

Figure 2. Government responsiveness and self-image.

If the two types of attitude are in fact correlated, we would expect not only that most of the cases would fall in the −− cell, but that more would fall in the ++ than in the remaining −+ and +− cells. Yet this may very well not be the case. The cell −+, "The government is responsive, but I am worthless . . . ," is where we would anticipate finding blacks subject to paternalistic modes of white-dominated government. And because Southern government is in fact extraordinarily unresponsive to blacks, a substantial number of Southern blacks presumably fall into the +− category as well as the −− category.

Whether there is or is not a greater incidence of lower political self-image among Southern blacks cannot simply be extrapolated from Northern versus Southern figures for black *nonparticipation* in politics. Because actual barriers to participation are encountered less often in the North, it is in the North that low political self-image is most likely to have a strong association with nonparticipation. Are Northern black feelings of political inefficaciousness substantially greater than among whites, once standard explanatory social and economic variables have been accounted for? If so, what if anything is there in the political system itself that can explain this? Is the system actually less responsive to blacks although not as severely so as in the South? Or is political cynicism somewhat independent of the way the political system actually works? Is it simply a minor segment of an overall pessimism and low self-image engendered by an informal social order which is not so strikingly variant from North to South as is the political system?

MILITANCY, SELF-IMAGE, AND POLITICAL EFFECTIVENESS

The spectacular growth of black militancy has substantially altered the distribution of black attitudes towards both government and self, dramatically increasing the proportion of blacks whose position is, "I am worth a lot, but the government is not responsive." Significantly, the militant's behavioral conclusion is not nonparticipation, but instead intense and forceful participation in an attempt to overwhelm unresponsiveness.

Some intellectual apologists for black separatism concede

that deficiencies among many blacks in positive self-image, as well as more tangible deficiencies in education, skills, and financial capital, necessitated the separation in order to gain time and experience and confidence. But the mood among virtually all militant blacks is one which rejects any such concession, and is boisterously self-confident in tone.[11]

The interrelationships among positive self-image, white frustrating of black aspirations, and physical violence are extremely complex. Some blacks use violence instrumentally (that is, *politically*); to force the governing system to respond, or to advance their own personal lives as political leaders in some calculated way. For others, violence is (or is rationalized as) a physical expression of positive self-image: the black equivalent of the "never again!" mentality of the Jewish Defense League. For still others, violence provides catharsis; it is done because it feels so good, regardless of any consequences which may follow. Violence perpetrated in order to induce catharsis comes closest to being the classical behavior predicted by Freudian psychological theorists, linking aggression to frustration.[12] Of course, some violence (such as rioting and looting) is done just because it is fun.[13]

The classical option of seeking relief from frustration through aggression is chosen either when conditions are so utterly desperate that even the possibility of death as a consequence is accepted; or—quite at the other end of the spectrum—the penalties seem rather remote. The most aggressive blacks in this era tend to fit the latter category more than the former. They tend to be relatively safe from economic and social sanctions, and to a lesser extent from physical sanctions. They are also young, relatively unencumbered by decades of repeated rebuff, insult, and injury.[14] Some blacks, however, are clearly in the utter desperation category, both in terms of their objective condition and their state of mind.

AGENTS OF POLITICAL SOCIALIZATION

On his way to adulthood, a black child or adolescent has to *learn* that the political system is or is not responsive, or that legitimate or illegitimate activity by him will have effects. He learns from a variety of socializing "agents," among them being the schools, his family, the media, and his peers. Solid research on the contributions of each of these agents to black learning of political facts and feelings is almost nonexistent, partly because of resistance by influential blacks to social science scholarship.

The schools

Most studies of the relationship between amount of education and levels of political activity or interests have compared adults who in the past accumulated a lot of education against those who in the past accumulated very little. It is precarious to attribute the greater amounts of interest and participation on the part of highly educated people to whatever it is that happened to them during the additional years of schooling they had. Additional schooling and political interest may, in fact, simply both be dependently related to a common antecedent independent variable; namely, social status of the family.

Nevertheless, it is worth reviewing the cross-sectional findings, since some material on blacks exists in this form. In the late 1950s this author studied adult black political participation in Durham and Winston-Salem, North Carolina, hypothesizing among other things the following: that while the general phenomenon of increasing participation with more education completed would hold for both races, the rate of increase for whites would be greater than it would be for blacks. While those with high-school educations in both races would generally participate at a greater rate than those with only grade-school educations, the increase in participation from grade school to high school would be markedly greater for whites than for blacks.

This hypothesis was only imperfectly demonstrated by the actual data. Excluding for the moment those whites and blacks who had attended college, black and white adults with up to a ninth-grade education kept pace with one another pretty well; that is, participation by those who attended junior high school exceeded the levels of participation of those who only reached grade school at best by roughly the same amount for each race. Black adults participated at a lower level than whites, to be sure, but the race differential remained essentially constant with the increase of education through junior high school. The hypothesis was borne out for those who attended junior and senior high school, however, with whites showing a greater gain in amount of political participation from junior to senior high school than blacks, even though all, of course, increased in participation to some extent.

The most striking data were for the college educated. In Durham, black adults with college educations participated at a greater rate than whites of college education. In Winston-Salem, the gap in participation levels at the high-school level was essentially closed at the college level; that is, college-educated whites and blacks participated at essentially the same level.[15]

Now let us assume for a moment that these cross-sectional data also hold true in the longitudinal sense, that an individual black proceeding through school "loses ground" to his white counterpart through high school, but then, after having gone to college, he gains sufficient ground to draw even or to go ahead in amount of participation in politics.[16] Since the family environment presumably has not changed during all these school years and since there has been no change of social status, no change in parental levels of participation and interest, etc., it would be fair to attribute the increases noticed over the span of school years to the school experience itself.

Giving the schools their due (or more than their due), then, what kind of explanation can we suggest for the greater proclivity for participation among whites than among blacks? One possibility is that, on the whole, black school children simply learn fewer political facts per year than whites and so are less capable, less willing, and less interested in participation by the time they come of age. A second possible explanation contemplates the school experience as a broader contributor to socialization:

> *In this sense, schooling is seen as a means of inculcating persons growing into adulthood with patterns of expected behavior. Thus school becomes along with other major institutions of society, a teacher of norms, a preparation ground for "real life." In the political sphere, schooling usually involves "learning to be good citizens." The political ideas current in the world are rarely presented in strictly analytical contexts. Instead, the inherent superiority of democracy over totalitarianism is alleged, and the consequent moral imperatives are derived from the comparison—for example, "Everyone ought to vote. . . . "*[17]

Taking the school experience in this spirit, one would normally expect the inculcation of the democratic creed to increase proclivity toward participation. For blacks, however, the total school experience may involve much more than simply learning the democratic creed. There are indeed other norms that blacks may perhaps learn in school, especially in those schools which are all or almost entirely black.

> *If schooling is primarily a matter of socialization, it might be hypothesized that Negro political participation rates would tend to decrease as the number of years of education increases, reasoning thus: The Southern Negro public school system (or possibly all black schools anywhere) tends to inculcate Negroes with the prevailing norm of political behavior for that race; namely, noninvolvement. The greater the exposure to this indoctrination (that is, the longer a Negro stays in school), the less the tendency to participate in politics.*[18]

The idea expressed here is not literally that we would expect to find proclivity toward participation actually decreasing as the black child proceeds through school but that this particular factor would tend to dampen the rate of increase that might otherwise apply (as with white children).

This second line of speculation makes some rather gross assumptions about what black children actually experience in school. Do all-black schools consciously or unconsciously inculcate resignation, apathy, or cynicism about politics? If so, is this more true in Southern black schools than in all-black schools in the North? Are reservations about the democratic creed which could lead to eventual apathy or cynicism actively stated by teachers, or are they simply observed by students who see teachers shy away from controversial issues and from active participation in sit-ins or street demonstrations? Does even the black child in the predominantly white school in the North learn by indirection that his participation in politics is not especially treasured? Answers are lacking and needed.

What we do know from direct empirical research is that in one Northern city, through the early school years, blacks show cognitive lag and affective difference within a closely circumscribed substantive area (the Greenberg Philadelphia study). But no evidence is given linking these findings to school experiences as distinct from life experiences in general at the various young ages. Preliminary findings from the massive, benchmark study of high school seniors by M. Kent Jennings and the Survey Research Center show much more clearly that the specific "civic education" experiences blacks have in high school measurably increase their positive affect towards the political system—an effect which does not occur for whites by this time of their educational careers.[19]

The family

Earlier the observation was made that much of the process of political socialization occurs during the school years. One could as well say, during the family years, because the two periods occur roughly at the same stage of life. If we who are concerned with the nature of the political education blacks are receiving, formally or informally, expect to alter the course of things, we must face the grim prospect of ascertaining what the parameters are within which we can operate. Realistically, this means determining what contribution the school experience makes to political socialization, as against the contribution of the family or other aspects of the environment. Unfortunately, the school contribution is rather small, judging from what evidence is currently available. While there are certainly "school effects"—an increased amount of political information that can be traced to school experience, an increased movement away from the political attitudes of one's family, etc.—these effects are still only marginal.

By contrast, the "family effect" is striking. The extraordinary perseverance of loyalty to a given political party within families is well documented. Hyman reports a correlation of 0.9 between party preference of parents and children.[20] Party loyalty persists in great strength even in the face of conflict with, for example, social class. Martin Levin demonstrates that when family political-party preference is in conflict with the social or occupational status of the family, it is the family tradition rather than the occupational class which overwhelmingly carries the day. He reports that of high school students whose parents were both Republicans and whose fathers were of blue-collar occupational stratum, 91 percent considered themselves

Republican. And again where occupational stratum and parents' political preference were in conflict—that is, where the parents were white collar but also Democrats—the children were overwhelmingly Democratic. In this case, 75 percent were Democratic.[21]

It is not only party preference which carries over from parents to children. There is also a strong correlation between the ideological positions of children and their parents. In this case, correlation is of the order of 0.5 rather than 0.9, according to Hyman. But still a rather substantial amount of rates of participation in politics remains roughly the same from parents to children.[22] Given evidence of this strength, it should not come as any surprise that the amount of political participation also tends to be similar within families.

In talking about parent-child political consistency, it is somewhat hazardous to lump both parents together. Of course, there is a great deal of consistency there as well. In the case of conflict between the parents regarding issues, the parent more likely to be followed depends in part on the specific issue area. But on the whole, the tradition of "politics is a man's game" has its effect, both by making the father the primary guidepost for the child's development of issue positions and for bringing the wife into line, should she deviate.[23]

These are figures for whites. One must consider whether family life in black America is significantly different from that in white America and whether any such differences that do exist have political relevance. The observation that black family life is more matriarchal than that of white comes to mind immediately. (No claim is made that black families are predominantly matriarchal, but only that more of them are than among white families.) If the matriarchy proposition is valid, we would expect greater influence of the mother of the household on political attitudes, feelings about political activity, learning of political facts, and development of political skills. We should not immediately jump to the supposition that this would entail lower levels of information, activity, and interest just because American women as a whole tend to be lower in all of these than men.[24] If, indeed, the black woman is more likely to assume male roles in the family in general, there is no a priori reason to believe that she does not also assume the typical male political role in the family.

In this author's North Carolina study, he presented some indirect evidence that in fact the black female does not pick up the role of political propagator within the family. A smaller observed difference in participation between black males and females than among white males and females was attributable to a drop in male participation rather than an increase in female participation.[25] If black women play a greater part in political socialization of their children than white women do for theirs, then the effect on the children would be lower levels of information, interest, participation, and skill.

There are several bases for the matriarchy theory. Some make reference to experience as slaves and even earlier as African tribal members.[26] Others convincingly point to the influence of the mother because of her earning power relative to the father in many black families. Perhaps most important is the larger incidence among black families of "broken homes"–broken through divorce, separation, lack of husband, and the husband's having moved North to earn more income. It is interesting to speculate what effect absence of a father has on the early political socialization of black children. Thinking again of the findings of Greenstein and Easton and Hess, does this absence make it more difficult to develop images of what political figures are like? Does a hypothesized greater difficulty in developing a strong, favorable image of prominent officeholders lead to greater cynicism in later life toward politicians? This line of thinking assumes that the early correspondence between an image of one's father and of the President or other officeholder, noted earlier, is not simply two instances of a generalized reaction to prominent males but is rather a specific projection of feelings about one type of male, the father, toward another who seems to the small child in some ways to be similar.

The crucial agents: mass media and peers

Black militancy, so heavily a movement of the young, has created a generation gap in political attitudes and behavior among blacks which dwarfs the more publicized gap associated with middle-class whites. Both its youthful aspect and the suddenness with which it has arisen point to a sharp discontinuity in the socialization process, conceived as the transmittal of values (among other things) from one generation to the next. The schools and the family are basically agents of smooth transmittal. Sudden discontinuity is typically generated by other agents, ones less encumbered by the inertia of tradition and more susceptible to influence by idiosyncratic catalytic events.

Specific events helped trigger the maturation and diffusion of black militancy. Over a very short time span, blacks saw white college students and their allies try to link the civil-rights movement to their own left-wing ideology and their increasing preoccupation with the Vietnam War. Soon after, they witnessed the virtual abandonment of civil rights by white liberals. And finally they saw and felt in the most profound, traumatic sense the murder of Martin Luther King.

Television faithfully documented and broadcast these events and trends, delivering the ugly news . . . and also instructing blacks, in effect, about new modes of social and political behavior.

But the most essential medium for the diffusion of new, militant expression of both political objectives and self-pride for blacks has been peers. Political and racial indoctrination has spread largely by means of oral and written communication among high-school and college-age peers, in friendship cliques, in clubs and gangs, and in an impressive array of formal organizations such as campus Afro-American Societies.

Schools, of course, have something to do with peer group political socialization in that they provide a daily reason for hundreds or thousands of peers to congregate at one location. They facilitate the diffusion of opinion among peers;[27] they often bring blacks into contact with the "enemy," permitting the use (by both races) of new political rhetoric and actions against each other on an experimental basis; and in similar fashion, they offer blacks (and others, but in these times especially blacks) program, policy, and personnel issues on which to test out and develop their political "muscles."[28]

SEEKING THE BOUNDARIES

Blacks of all ages, but particularly the young, are testing out white America, seeking specific political, economic, and social goals; cultivating a personal sense of worth and influence; and trying to map out the boundaries for acceptable political action that American society now sets for them. To black amazement (in what is basically a repressive culture when it comes to race), white Americans have delineated very few limits. Extremist blacks, for example, find that a surprising number of white liberals come to their side even in the most outrageous circumstances, including murder. In some cities, political actions by blacks are tolerated by police and politicians which are not tolerated when perpetrated by whites.

This anomic response is not likely to persist indefinitely. One can only hope that when the boundaries of political legitimacy do begin to take clearer shape, it will be because nonblacks have decided to insist on rigorous enforcement of their democratic norms as Americans, not their racial norms as whites.

Despite separatist objection, there is good cause for whites to intervene as best they can in the political socialization process of blacks. It is to everyone's advantage to explore the meaning of political legitimacy in a democratic society—and to have whites work with blacks to maximize their position in society by the most skillful use of legitimate political action. Perhaps the greatest contribution educators can make to school-age blacks who will be tomorrow's adult citizens is to reorient their thinking about the development and use of political strategy. This means spelling out with approval the various techniques of bargaining, forced demands, concession, and occasional retreat that are used by politically successful subgroups in our society. It means being candid on two scores when dealing with heterogeneous groups of students in the classroom—candid about the probable maximum of political potential that a given subgroup could have (just how successful blacks can expect to be, given their total resources of numbers, money, effort, education, and so forth) and candid about the kinds of political techniques that are in fact being used currently or may be used in the reasonable near future.

Political activity in this day and age, after all, involves not only voting, contributing money, and writing letters to congressmen. It sometimes involves street demonstrations and civil disobedience. These need talking out in the classroom, too, not in normative terms but in terms of strategies which sometimes succeed or fail because they tread so close to the border of normatively acceptable political behavior. Our success in drawing black youth into active and thoughtful citizenship may ultimately rest on candid and imaginative assessment of a new politics blacks have helped create.

REFERENCES

1. The typical limitation to whites in research designs concerned with political socialization seems to be a consequence of a desire to simplify research strategy rather than to overlook an important variable.
2. Herbert H. Hyman, *Political Socialization*, (New York: Free Press, 1959).
3. Edward S. Greenberg, "Children and the Political Community: A Comparison Across Racial Lines,"*Canadian Journal of Political Science* 2, no. 4 (December 1969), pp. 471-492.
4. In an unpublished study, Frederick W. Frey of M. I. T. observed increasing cynicism among senior high school students in a Boston suburb as compared with junior high school students.
5. David Easton and Robert D. Hess, "The Child's Changing Image of the President," *Public Opinion Quarterly* 24 (1960), pp. 632-644. For a full exposition of childhood political learning, see: R. Hess and Judith V. Torney, *The Development of Political Attitudes in Children,* (Chicago: Aldine, 1967).
6. Fred I. Greenstein, "More on Children's Images of the President," *Public Opinion Quarterly* 25 (1961), pp. 648-654.
7. Greenberg, loc. cit.
8. At least two of the items in the Survey Research Center's Political Efficacy Scale are ambiguous between these two attitudes. For a discussion of the relationships between a sense of political efficacy and political behavior, see especially Angus Campbell et al., *The American Voter* (New York: Wiley, 1960).
9. Donald R. Matthews and James W. Prothro ingeniously demonstrate the substantial contribution of such political factors (as against social and economic) as formal voter requirements, state factional systems, existence of racial political organizations, party competition, and racial violence to one form of political activity, registering. *Negroes and the New Southern Politics* (New York: Harcourt, Brace & World, 1966), Chap. 6.

10. Levin encounters a parallel difficulty with the concept of "political alienation." Alienation is postulated as inducing four distinctly different behaviors: rational activism, withdrawal, projection, and charismatic identification. Murray B. Levin, *The Alienated Voter: Politics in Boston* (New York: Holt, 1960).

11. Benjamin F. Scott provides a compassionate account of the rise of black pride in *The Coming of the Black Man* (Boston: Beacon Press, 1968).

12. John Dollard et al., *Frustration and Aggression* (New Haven: Yale University, 1939).

13. Edward C. Banfield, *The Unheavenly City* (Boston: Little, Brown, 1970), Chap. 9.

14. *Report of the National Advisory Commission on Civil Disorders* (Washington, D.C.: Government Printing Office). The *Report* characterizes the typical black rioter as one who "takes great pride in his race and believes that in some respects Negroes are superior to whites. . . . " (p. 128).

15. Bradbury Seasholes, "Negro Political Participation in Two North Carolina Cities," Ph.D. diss. Chapel Hill: University of North Carolina, Department of Political Science (1961), pp. 128-147.

16. There is obviously a problem here in that we can't reasonably talk about persons' level of *political participation* during their actual years of schooling, when at that age many forms of political participation are legally not open to them. It would be better to think of increasing the students' sense of favorable attitudes toward participation, of "If I could participate, I certainly would."

17. Seasholes, p. 84.

18. Ibid., pp. 86-87.

19. Kenneth P. Langton and M. Kent Jennings, "Political Socialization and the High School Civics Curriculum in the United States," *American Political Science Review* 62 (1968), pp. 859-867.

20. Hyman, op. cit., p. 74. A high correlation is helped, of course, by having only two parties to choose from.

21. Martin L. Levin, "Social Climates and Political Socialization," *Public Opinion Quarterly* 25 (1961), pp. 596-606.
22. Hyman, op. cit., p. 72.
23. Fred I. Greenstein, "Sex-related Political Differences in Childhood," *Journal of Politics,* 23 (1962), pp. 353-357. See also: Hyman, op. cit., pp. 83-84.
24. Lane, op. cit., pp. 209-215.
25. Seasholes, pp. 128-147.
26. See especially: Melville J. Herskovits, *The Myth of the Negro Past* (New York: Harper & Row, 1941); and E. Franklin Frazier, *The Negro Family* (New York: Macmillan, 1939). Herskovits was particularly sanguine about American Negroes' ties with Africa. See also: Harold R. Isaacs, *The New World of Negro Americans* (New York: John Day, 1963), pp. 109-110.
27. Diffusion requires many other favorable statuses besides proximity, of course. See: Everett M. Rogers, *Diffusion of Innovations* (New York: Free Press, 1962), pp. 124-143.
28. For a review of the extent to which racial disorders contribute to high school disruptions of all varieties, see: Shelley R. Garrett and William G. Nowlin, Jr., *"High School Unrest: The National Parameters,"* mimeo (Medford, Mass.: Tufts University, Lincoln Filene Center for Citizenship and Public Affairs, 1970); and Stephen K. Bailey, *Disruption in Urban Secondary Schools* (Washington: National Association of Secondary School Principals, 1970).

5

Enhancing the Black Self-Concept through Literature

Nancy L. Arnez

The purpose of this paper is threefold. One purpose is to reveal some new research about the self-concept of black children. Another purpose is to show how the ingrained biases of white critics impede the use of black literature in the schools to enhance the positive self-concept of black children. A further purpose is to suggest the use of new revolutionary black literature as a means of enhancing this positive self-concept.

The significance of identity problems for the young black child has been researched by a number of investigators. The prototype for later studies was carried out by Kenneth and Mamie Clark in the midforties.[1] Other studies followed: Goodman 1952,[2] Trager and Yarrow, 1952,[3] Landreth and Johnson 1953,[4] Morland 1958,[5] Stevenson and Stewart, 1958.[6] The findings converge in indicating that racial recognition by both white and black children appeared by the third year and sharply increased thereafter. Of special import in all these studies was the tendency of black children to prefer white skin, white dolls, and white friends. They often identified themselves as white or were reluctant to acknowledge that they were black. Related to this was the fact that young children of both races assigned poorer houses and less desirable roles to black dolls. The literature also indicates that some residuals of this negative self-concept were found in older black children.

These identity problems, the literature suggests, are linked with problems of self-valuation (self-esteem). Black people's assignment to second-class status together with the white racists' insistence on innate inferiority, we are further told, no doubt have been instrumental in creating in black people doubts concerning their own worth. Consciously or unconsciously, the literature emphasizes, many black people have accepted in part these assertions of their inferiority.

MORE RECENT RESEARCH ON BLACK SELF-CONCEPT

Four recent studies reported below reveal the opposite position. The Larson, Olson, Totdahl, and Jensen study done on white and black kindergarten children in five inner-city and two outer-city schools, reveals that although black children more often incorrectly identified themselves racially than white children, they showed no significant preference for either race in their positive and negative role assignments.[7] This latter fact is of particular significance since most, if not all, of the studies on racial awareness and self-concept among children done up until 1966 report that black children generally give negative role assignments to black dolls or pictures.

A second study pointing toward a positive self-concept in black children is the Georgeoff study of fourth-grade white and black children. The three groups used in the study were divided, by random, into Group A, including ten classes of black and white children from the same neighborhood; Group B, with nine classes of black and white children from different neighborhoods; and Group C, the control group, which included seven classes of black and white children from the same neighborhood.

Only the experimental classes (Groups A and B) were taught the unit on "The American Negro." At the end of the unit the Piers-Harris Measure of Self-Concept was administered to these classes.

Analysis of the data reveal that for the black children no significant difference exists between Groups A and B or between Groups B and C, but a significant difference at the .05 level did exist for Groups A and C. This study, then, reveals that the self-concept of the black children in the experimental groups who studied the unit on the history and culture of the Negro was significantly improved.[8]

Another of the newer studies on black self-concept investigated the change in fifth-grade black students' pride and self-concept after exposure to black studies. The experimental subjects were obtained from two segregated black and two integrated classes. The control subjects were obtained from one segregated black class and two integrated classes. The control classes were not presented with black studies. The self-concept instrument used was the Self-Concept and Motivation Inventory developed by Farrah, Milchus, and Teitz. In addition a "Black People" Semantic Differential was used to assess black students' racial pride. Results of the study revealed that black students exposed to black studies made significantly greater gains on the "Black People" Semantic Differential than did control students. Further, results of the self-concept instrument revealed a significant difference in favor of experimental black subjects from the two segregated black classes as compared to the control subjects from segregated black classes. Also, black subjects in integrated classes had more positive attitudes than black subjects in segregated schools. In summary, the results of the study revealed that black students who are provided with black studies have positive racial pride. The data indicate that black students do not have poor self-concepts as measured by the Self-Concept and Motivation Inventory.[9]

Hodgkins' and Stakenas' study done on effects of segregation on the development of the black self-concept, also rejects the negativism syndrome. This study was done on 142 black and 100 white subjects of high school and college age. Generally, the school and social life of the subjects were segregated. The self-concept measurement was drawn from the semantic differential as developed by Osgood plus the investigators' vocabulary from which 27 bipolar adjective items were developed and arranged in random fashion under self-situation concepts. Results of the study indicate a significant difference between black and white subjects existed in self-adjustment and self-assurance in the school situation, with blacks tending to score higher than whites.

An important consideration in this study was that self-perception is not dependent upon the values of the total society but upon a person's evaluation of his performance in terms of role expectations and the "significant others" in his life. These significant others are considered to be persons with whom he interacts and those who have the greatest direct social control over him such as members of his primary reference groups.[10]

The implications of the above studies are indeed significant to social scientists if they are to move away from reliance on now-obsolete findings of research done in the 40s, 50s, and early 60s which reveal a negative self-concept of black children, that is, if they are to take a fresh new look at the consequence of events like the Black Revolution on the attitudes of black people. This is extremely necessary if black people are not to be crippled psychologically by being chained to findings of studies done twenty, ten, or even five years ago before the rush of events connected with the Black Revolution. Adherence to old findings will be detrimental to the operation of educational programs, and to political and social aspects of life.

As most of the older studies indicate, it is common for white and black social scientists to write and reiterate that black people, in general, have a negative self-concept. Even if this were true, and it is not, it is a defeating frame of reference for a whole group of people. White Anglo-Saxon Protestants have never allowed the above to begin. White people have constantly, and emphatically denied it through their own writers of fiction, nonfiction, psychology, sociology, and politics, and they have done this for a very good reason. The fact of continued and widespread emphasis is in itself negating. Continued repetition will, and has in the case of black people, crystallize negative attitudes of teachers toward children and their home conditions, politicians toward their black constituents, psychologists' and psychiatrists' interpretations of data from their patients,[11] other social scientists toward their subjects,[12] and literary writers toward their protagonists. For example, Wright began from the premise that what whites thought of blacks was more important than what blacks thought of themselves. Thus, Bigger Thomas was presented as an almost subhuman man that Wright designed to shock whites into ending the circumstances which produced such a creature.[13]

In order to counteract the above, blacks must take the

position of black narcissism, the position that black people love themselves, that in spite of conscious efforts to cause self-hatred by the European oppressor, black people have continued to personally feel good about themselves. Further, a black cultural and ethnic chauvinism is necessary to balance the pendulum of negativism toward blacks. Because others have thought blacks inferior and have hammered this falsehood into themselves and into their children, blacks must circumvent these efforts to psychologically maim them. This maiming has not really happened and certainly will not occur if black people consciously, conscientiously, and continuously parry efforts to make it so. First, though, blacks must recognize these efforts to negate them, subtle though they be. Therefore, let us focus momentarily on some of these efforts which are with us even today.

SOME EFFORTS TO NEGATE BLACK PEOPLE

Foremost is the effort to make black people believe that they have a negative self-concept by repeatedly writing about it. This negativism is revealed in studies on the black self-concept from Goodman (1946) to Morland (1966). Note this twenty-year emphasis in social science literature[14] on how black people hate themselves.

Another effort to negate black people is the way the language of whites serves to denigrate blackness. As Ossie Davis pointed out, a superficial examination of Roget's *Thesaurus of the English Language*[15] reveals 120 synonyms for blackness. They include "blot," "blotch," "slight," "smut," "smug," "sully," "begrime," "soot," "becloud," "obsure," "dingy," "murky," "threatening," "frowning," "foreboding," "forbidding," "sinister," "baneful," "dismal," "wicked," "malignant," "deadly," "unclean," "dirty," "unwashed," and "foul." Included in the same listing were words such as "Negro," "Negress," "nigger," and "darkey." On the other hand Davis found 134 synonyms for the word "white," almost all of them with favorable connotations expressed in such words as "purity," "cleanness," "immaculateness," "bright," "shiny," "clean," "clear," "chaste," "unblemished," "innocent," "just," "straightforward," "fair," and "genuine."

Podair[16] emphasizes Davis' point by stressing that there is a link between language patterns and race relations. The English language stereotypes black people, thereby intensifying race prejudice. It shapes ideas, concepts, and attitudes about people. Podair insists that the points of view expressed in "dark horse," "black," "black sheep," and "black and white," are difficult to counteract and, therefore, should be of import to the social scientists working in the area of race relations.

Margaret Burroughs' poem "What Shall I Tell My Children Who Are Black" further clarifies the subtle psychological effort to make black children hate themselves through the use of standard English:

They are faced with abhorrence of everything that is black.
The night is black and so is the boogyman.
Villains are black with black hearts.
A black cow gives no milk. A black hen lays no eggs.
Bad news comes bordered in black, mourning clothes black,
Storm clouds, black, black is evil
And evil is black and devils food is black . . .

What shall I tell my dear ones raised in a white world.
A place where white has been made to represent
All that is good and pure and fine and decent,
Where clouds are white and dolls, and heaven
Surely is a white, white place with angels
Robed in white, and cotton candy and ice cream
And mild and ruffled Sunday dresses
And dream houses and long sleek Cadillacs
And angel's food is white . . . all, all . . . white.[17]

Pentecoste points to the dilemma of the television industry, " . . . to satisfy the black demand for better dramatic parts and at the same time manage to offend as few whites as possible."[18] The four television shows analyzed were "The Outcasts," "Julia," "Land of the Giants," and "I Spy." Pentecoste's analysis reveals that the slave in "The Outcasts" appeals to black audiences because his role is that of a splendid black physical specimen who fights with and even kills some white men. The appeal to whites is that this former slave is treated like an overgrown child by his white owner, and the only whites killed by him are those who deserve killing in an effort to protect the white man.

Here we again have, as Pentecoste puts it, "The black body losing out to the white mind. . . ."[19]

"Julia," a light situation comedy revolving around the experiences of a young black widow and her son excludes entirely a viable black male character. When a black male is occasionally used, he is depicted as a "buffoon or small-time grifter." The only white man of consequence in the drama is old and, theoretically, cannot be viewed as a sex symbol.

Pentecoste also analyzes the black actor in "Land of the Giants." In this drama, as in "Mission Impossible" and "Mod Squad," we see that all of the black men are depicted as superhumans. For instance, in "Mission Impossible," we see a mechanical genius. In "Mod Squad" our view is focused upon a black man with a police-dog loyalty and obedience to orders. In "Land of the Giants," we have a living computer, and in "I Spy," a linguist. All, as Pentecoste implies, must be "superblacks" in order to be acceptable to ordinary television-viewing whites.[20]

Cecil M. Brown further emphasizes that television is really being used as a weapon against the developing black consciousness which undergirds a positive concept of self in black people. He says that "The media is counterrevolutionary, that is, it is against destroying the white viewer whom it pacifies—because the moment protest is introduced into the media it ceases being protest and becomes 'the news', 'a new T.V. show about spooks.' "[21]

Another example is the effort to redirect black self-determination into acceptable white standard thinking and action as the following excerpt shows:

> *Two years ago, Stokely Carmichael shouted "Black Power" and white America heaped vituperation on this young, beautiful man. In July 1968 James A. Linen, President of* Time, *Inc. speaking to the Urban League Convention in New Orleans said: "Of course, I'm for Black Power." There was, of course, no anger in the press, no waspish abuse, and no attempts to "understand." What was taboo two years ago, was now accepted by the white press as a fact of life. The outcry against and the subsequent acceptance of Black Power deserves attention for only the most doltish mind*

can fail to see what is operating here. Quite simply, it is this: When blacks promulgate a new idea, a new life style, it must be beaten down, or made to look obscene, or threatening to the nation, or lacking in intellectual content. When that fails, when blacks make it unmistakably clear that they have a brain which they intend to use, that they have their own heroes whom they embrace, then white America regroups, rethinks, and attempts to smother with affection that which it once rejected. This tableau has been played before, and now, it is going to be played with increasing frequency. For what is current in the land is a refusal on the part of many blacks to continue to let white people define them, label them, and so manipulate them with ease.[22]

The need for various civil rights acts shows that blacks cannot be protected under the original law of the land. A corollary to this is that the discriminatory clauses have never been erased from the Constitution and/or the Declaration of Independence, both documents reserved primarily for white men.

These examples illustrate some of the means by which whites have tried to destroy the dignity of black people. In order for this destruction not to occur, black people must be ever aware of these and the various other methods which have been and still are being used by the oppressor to try to make them hate themselves and their group. For it is only through such constant vigilance that black people can ward off the dangers of this psychological warfare being waged to weaken their resolve to be self-determining and self-directed.

BLACK POSITIVE SELF-CONCEPT

After recognizing the efforts of white America to psychologically paralyze blacks in an effort to maintain the white norms of society, the next step is to counteract this effort by focusing on ways to dignify blackness and to emphasize in writings and speeches the positive values of being black. Black people need not analyze and articulate the misery of black people anymore. Wright, Baldwin,

Killens, and many others have done that quite well already. Now, all energies must be expended on enhancing the positive self-concept of black people. It is as Coombs says:

> *Black health, I submit, now rests in black hands. . . . What we must understand is that no indoctrination is too extreme to defeat the racist purpose. . . . When CBS televised a class from the Freedom Library Day School in Philadelphia for Part One of, "Of Black America," the nation saw a glimpse of this indoctrination, and whites cringed and labelled the children's posture "fascist." But I know of no blacks who did not rejoice in that sequence. For they know that a terrifyingly bad school system would attempt to destroy their children, would try and tell them that academic courses were too rough for them, that they were expected to take shop courses, expected to live again the same mediocre life of their parents. So black parents rejoiced. They glimpsed the possibility of these young boys and girls growing with a sense of who they were, and what they were all about, and having at last the apparatus to deal with those who attempt to deny them their hope of life and who would assign them to the garbage heap, and gloat at their suffering.*[23]

The day Rosa Parks thought more of her tired feet than of her safety was the beginning of the display of mass positive self-concept in black people. This was the day black people decided to reveal their positive self-concepts. For this fact like many others was hidden behind the "shucking" and "jiving" antics of blacks as protection against the barbarity of whites. What followed, of course, was an affirmation of this mass positiveness through the sit-ins, the Freedom Rides, the protest marches, and finally the riots. Rosa Parks opened the springs of self-love in black people hidden these many years from the prying eyes and ears of whites. Hidden for fear of retaliatory efforts designed to make black people believe, as white scientists and social scientists had failed to do, in their hatred of self and of each other.

Actually, blacks have always had a mass positive self concept. For instance, the Uncle Tom-ing showed a love of self, for it enabled blacks to continue to live. The severity of black mothers with their sons revealed this love of self to perceptive people, who understood that when black mothers forced their sons to be Uncle Toms it was done in order to "raise them." The humor of slavery days in the folk tales about Old Marster and John suggest survival techniques used by the slave which most whites, including Joel Chandler Harris, never understood. The use of these and other survival techniques during slavery and after suggest the love of self of black people. Since a hatred of self could have led to an almost total destruction of the black race in America, the existence of thirty million blacks (including those black males the census is never able to locate)[24] in this country today suggests a love of self in black people.

Although Kardiner and Ovesey[25] declared that self-hatred is demonstrated when black males use hair-processing techniques to deny a black attribute, it is the reverse that is true. This hair-processing technique is used to heighten masculinity. It is, as Keil says:

> *The hair-processing techniques that Abrahams finds "reminiscent of the handkerchief typing of Southern Mammies" are designed to heighten masculinity. Backstage at the Regal Theatre in Chicago "process rags" are everywhere in evidence among the male performers, the same performers who put the women in the audience into states that border on the ecstatic. Prettiness (wavy hair, manicured nails, frilly shirts, flashy jackets) plus strength, tender but tough—this is the style that many Negro women find irresistible. A blues singer is not unconsciously mimicking Elvis Presley's hairdo (the opposite may be true) or Aunt Jemima's when he straightens his hair and keeps it in place with a kerchief. He is enhancing his sex appeal—nothing more.*[26]

The Chicago West Side black youths, advised by the faculty of the Center for Inner-City Studies, have never exhibited self- or group-hatred. In fact, these East Garfield Park youths are self-determined, self-directed, and show both race and individual love and pride. Some time ago, these youths became disillusioned with the public school system and left high school before graduating. Because of their disillusionment, they decided to establish an alternative school system for the people in their community. Thus, they negotiated with some priests to occupy an empty convent and school on a one acre lot in the area. Both the young men and women in the area proceeded to clean up the convent and school in preparation for the opening of their own community school. The priests then reneged on their verbal agreement and sent the police to evict the young people from the facility and land. Many of the young people were injured in the fight with the police, but, nevertheless, they remained in the building. Following this incident an Open House Affair was held to apprise the community of their intention to conduct a relevant school program for children and adults. Notables such as the Reverend C. T. Vivian and Mr. Augustus Savage, a politician and editor, attended along with many faculty members and students from the Center for Inner-City Studies. The success of this affair did not deter the priests from again sending the police into the convent to evict the youths. This occasion also resulted in numerous arrests and injured youths. Therefore, it was after this second invasion that the faculty of the Center for Inner-City Studies advised the youths to move to a structure offered by another religious group. This they did and continued plans to set up their alternative school system. Recently, the director of the Center for Inner-City Studies received a communication from the group of young men indicating that they had just filed their charter for a community school whose purpose is:

> *To provide educational facilities for all age groups in the community with specific emphasis on health, recreation, and training programs. These programs will be designed by the youth with full participation of the community.*[27]

Furthermore, the communication states:

> *The Board of Directors will serve in an advisory role with all final decisions being made by the Executive Committee of Pettis College.*[28]

The significant point about the above is that the president of unknown Pettis College, Inc. is a young black man who is about twenty years old. This, in itself, speaks to the issue of the very positive self-image of young blacks. But even more than the age factor, the entire situation shows the self-determination and self-direction of a group of young black men.

Further proof of the positive self-image of young black men is the thrust which the Black Panther Party, the Black Disciples, and the Black P. Stone Rangers are making in Chicago and elsewhere with their breakfast programs and/or free neighborhood medical clinics which give comprehensive health services to the needy. Many doctors and nurses volunteer their services six days a week from 6 P.M. to 9 P.M. and on Sundays from 10 A.M. to 5:30 P.M., so that adults, with or without children, will not miss a day of work. Specialists include pediatricians, gynecologists, obstetricians, eye-ear-nose and throat specialists, laboratory technicians, x-ray specialists, dentists, and psychiatrists.[29] The several breakfast programs are generally operated between 7:45 A.M. and 8:45 A.M., Monday through Friday, in several poverty areas. The food is filling and well-balanced. Also, the various youth organizations in Chicago joined with the Coalition for United Community Action in closing down construction sites until negotiation was completed. This negotiation with the Mayor led to 4,000 jobs for the black community in the building-trades industry and the establishment of training programs to prepare blacks to go into trades as electricians, plumbers, plasterers, and heavy-duty equipment operators.[30] All of the above discussion supports the thesis that black youth as a whole do indeed have positive self-concepts, and are self-determining and self-directed.

Recently, too, there are new insights into the role of the black mothers in their effect on the masculinity of the black male. The negativism of the Moynihan Report of 1965 on the black family is again being challenged in a recent discussion of black manhood. This report suggests that there is no lack of a male image and that, in addition, females also instruct male children in the expectations of the typical behavior of men, according to their experience of man-woman relationships.[31]

In summary, now that black people have rejected the theories of cultural deprivation and cultural disadvantagement, white-oriented social scientists are endeavoring to prove that black peoples' lack of success in America is due to their negative self-concept rather than to the racist nature of society and its exclusionary institutions and methods. The question then is, how do blacks counteract this new thrust in the psychological warfare now being waged against them? One way, of course, is through the use of literature by the new black writers in the schools. Much of the writing of these new writers reveals a positive feeling about blackness in general, and about the black life style in particular.

NEW BLACK LITERATURE

White critics

Literature is only one means by which black people can enhance the positive self-concept of black children despite white society's oppressive institutions; its colonialist methodology and its miasma of barbarity. But, in order to do this, there must be an oncoming surge of black critics born out of the same, or similar, experiences of black writers. A corollary to this must also be the destruction of the stranglehold on literary criticism by white writers who are deemed by the establishment to be more qualified to judge black writing than are black critics. In effect, black writers and black people must "put down" white critics such as Robert Bone, Edward Margolies, Herbert Hill, Leslie Fiedler, and David Littlejohn, who are considered to be experts on black writing. There is no possible way that whites can evaluate the black experience in or out of a literary context. They wear blinders because of their white perspective and the totally white quality of their lives. They can have but one approach to black literature. And that is to measure it in terms of the degree to which it reflects white life in America and white cultural values.

No longer can black writers be concerned with such standards as universal appeal of themes, white-denoted excellence in technique, control, imagery, style, and language. Black writing needs allusions to the black life style, experimentation with unique rhythms, different syntactical forms, and new orthography in tune with speech patterns of black people. We need to look at the inventiveness of vocabulary and the real-life character of events, creativeness of expression, and its focus on the humanness of man. What white man has the soul to tune in to the genius of black folk? Much of the negative criticism of black writing is born out of a one-sided view of something of value—a misunderstanding of the relationship between life and art. Our young black writers, most of whom are under thirty-five, challenge white criticism of the new black aesthetic. They believe in themselves, they have love for black people, and they use a unique creativeness to express this love.

Some white critics say that this new black writing is protest writing. Protest writing is directed toward white people; the new black writing is directed toward black people. As Kgositsile says, "Jones is not addressing white American or some ideal universal intellect or conscience. He is talking to his people about his condition, finally theirs, because he is his people."[32] He emphasizes this even more when he says, "The critics, of course, are the white murderers who want to control everything, including the black poet's creativity.[33]

Even when black people make clear the intent of their poetry, there is misunderstanding.[34] White critics cannot comprehend what a black writer is doing. Therefore, as Lee says:

> *We will determine standards of judgment and excellence and* no *white boys in the pages of the* Nation, *the* New Republic, Saturday Review, New York Times, *will direct or effect our efforts. Black people will direct us; direction will be a reciprocal process, shared between black people and black artists, and as sister Gloria House has said:*
>
> *You can't just go and ask anybody man.*
> *Cause, dig, they just don't know.*
> *They ain't - been where you go.*
> *EVERYDAY*
> *They ain't been through your neighborhood, man*
> *They can't give you no directions!*[35]

How can black people ignore the conditions of their lives and focus only on bees, birds, and trees? Black art must mirror the political and social life of black people beacuse the survival of black life depends on it.[36] The black artist shows black people a concept of self and points the direction for their survival.[37] In black journals we find a glorification of the existing black culture in a conscious effort to build a sense of identity from the past. The emphasis is on the use of sociopolitical themes, buttressed by an artistic-literary style of writing[38] devised by black artists and understood by black people.

Black writers for some time to come must, as our young black writers of today are doing, combine sociopolitical themes with the aesthetic. They must be the clamoring voices giving a "call to arms" to black people for unity and black awareness and black consciousness. They must continue to mythicise and symbolize the lives of black people. They must draw together through the vitality of their art, the fragmented, the distorted, and the maligned lives of black people. At present there can be no separation of art and life, and can there ever be, for what is art but life relived, revisited, through another form?

Innovative approaches to the use of literature

Teachers can enhance the positive self-concept of black children by using literature written by the young black writers. The emphasis should be on the younger writers such as Don Lee, Sonja Sanchez, Jewel Latimore, Mari Evans, Carolyn Rodgers and Etheridge Knight. Their works carry a revolutionary message, a message about the beauty of blackness, a message of nationhood. All ethnic groups have used their poets to sing the praises of their people. So too must black people, to indoctrinate black children into the belief that they are valuable human beings.

Unfortunately, young black writers have been more-or-less excluded from all publications but the *Black World*, which has featured them through the years. Hoyt Fuller, the editor, has a deep interest in exposing the creativity of young writers. Many have also been featured in single volumes devoted exclusively to their poetry by Dudley Randall, the publisher of Broadside Press, who is equally devoted to exposing the genius of black writers.

There are a few anthologies which include the works of one or two of the new black writers. One such is *Black Voices*, an anthology of fiction, poetry, autobiography, and criticism, edited by Abraham Chapman. The only poetry anthology to date in which two-thirds of its pages are devoted to the works of new black writers is *Black Poetry: A Supplement to Anthologies Which Exclude Black Poets,* edited by

Dudley Randall.[39] This collection contains poetry published from 1966 to 1969 and includes the black consciousness poetry and the black revolutionary poetry.

It is unfortunate that there are no anthologies devoted entirely or almost exclusively to new black short story writers, although John Henrik Clarke's *American Negro Short Stories* and Langston Hughes' *The Best Short Stories by Negro Writers* both contain the writings of a few new black writers. The best source of short stories revealing the black consciousness theme can be found in recent issues of the *Black World.*

Black Arts: An Anthology of Black Creations, edited by Ahmed Alhamisi and Harum Kofi Wangara[40] contains essays, articles, plays, short stories, sculpture, illustrations, drawings, graphics, fashions, photographs, poetry, and interviews. Any teacher of black arts will find this publication a valuable resource, since its focus is entirely upon the works of the younger black artists and writers.

New Plays From the Black Theatre, edited by Ed Bullins[41] is an anthology of eleven new plays by ten black writers. These plays, which dramatize the black experience, are being performed in black theatres in the black community for black people written by black writers, generally of the younger generation.

The writings of the older writers such as Hughes, Killens, Ellison, Brooks, McKay, Dunbar, Bontemps, Wright, Baldwin, Himes, and Petry should not be ignored. However, we must focus on the new themes to enhance the positive self-concept of black children. Young writers avoid the subtleties of the older writers, who feared that, if their messages were too clear, their works would never be published.

Black writers preserve the beauty of blackness by passing it down through the generations, by avoiding the homogenizing quality of integration as we know it today. They preserve and protect the concept of black ethnicity which, in itself, will strengthen and enhance the positive self-concept of black people.

This new writing enhances black heroes chosen by black people. It proclaims the greatness of their deeds as viewed by black people. And because many of these new black writers do not rely on white journals for publication or on white critics as the significant others to praise their efforts, their works, in black publications, hit home. The depth and potency of the artists' feelings are laid bare upon the pages. Furthermore, these new writers see black people as the significant ones in their lives and, therefore, they gear their writings toward black audiences. Black people who read these works are made further proud of their heritage and contributions to the world culture. Hence, their self-concepts are further enhanced.

There are several specific things that a teacher of literature must take into account when approaching his subject. For one, he must find promising instructional innovations in the use of new black revolutionary poetry. Much of the new poetry is not to be read–but to be recited. In addition, a recitation by the author gives the essence of meaning to the poem and establishes a bond between the author and the audience similar to that established between the preacher and his parishioners. Thus, every teacher of literature should have access to tapes of the new poets.[42]

The use of *Broadsides,* the single poster poems, which can focus the attention of the entire class on one poem at a time in written form is another way of enlivening the study of poetry and ensuring maximum literary identification. Among these masterpieces which can be obtained from Broadside Press[43] are "A Poem for Black Hearts," by LeRoi Jones; "Child's Nightmare," by Bobb Hamilton; "I Heard a Young Man Saying," by Julia Fields; "Sunny," by Naomi Long Madgett; "Race Results, U.S.A., 1966," by Sarah W. Fabio; "Back Again, Home," by Don L. Lee; "Black Madonna," by Harold Lawrence; "2 Poems for Black Relocation Centers," by Etheridge Knight; and "Black Unite," by Bahala Nkrumah.

Since there are so many selections on the market today, the teacher can use some help in the selection of materials to incorporate into her literature study. A few distinguished black journals are *Freedom Ways, Liberator, The Black Arts Magazine, Black Dialogue Magazine, Black Theatre, Black World, and Black Expressions.*[44]

For single references to moving, alive, and exciting black awareness poems, a teacher need but to select from Broadside publications which include: *Black Poetry: A Supplement to Anthologies Which Exclude Black Poets,* edited by Dudley Randall; *Black Pride,* by Don L. Lee; *For Malcolm: Poems on the Life and Death of Malcolm X,* edited by Dudley Randall and Margaret Burroughs; *The Rocks Cry Out,* by Beatrice M. Murphy and Nancy L. Arnez; *Think Black,* by Don L. Lee.[45]

The use of rock or pop songs as poetry will add spice to any classroom and will further constitute a way of enhancing the self-concept of black children. For example, as one listens to the song "Think" as sung by Aretha Franklin, it is easy to grasp that the song is a direct, frank portrayal of today's world. The song clearly states its message so that a teenager can understand the importance of freedom. "Think" contains rhythm, natural language, meaning, and strong feeling. Cherry A. Banks writes:

> *Songs such as James Brown's "I'm Black and I'm Proud," and Roberta Flack's "I Wish I Knew How It Felt To Be Free" express some of the pent-up feelings of the black people. The lyrics as well as the music of these songs reflect the black man's experience in this country. Today's rock songs by black artists abound with lyrics on social issues with which black children can easily relate.*[46]

Viewing, reading, and discussing plays written by younger black playwrights is another way of dealing with the black experience in a literature class. Plays such as LeRoi Jones' *Slave Ship*, about the inhumane voyages of the "middle passage"; Jimmy Garrett's *We Own the Night*, a drama of today's Black Revolution; Lonne Elder, III's *Ceremonies in Dark Old Men*, a family drama about black survival; and Marvin X's *Take Care of Business*, a play about the repudiation of the American Dream, are all realistic portrayals of the black experience.

Visual arts is another form that a teacher can use as a stimulant for literary expression. Black photographs speak forcefully of the black experience when their subject matter is the daily life activities of black people. The immediacy of the reproduction of the image can inspire quality expressions in writing of despair, agony, love, happiness, and other emotions.

A study of the drawings of Charles White[47] could be incorporated into any literature study which focuses on black themes, for the lines and shadings of White's works communicate the suffering, joy, and dignity of life of black people. His paintings sing a story to the viewer. One has but to see White's "Woman of Sorrow" and "Oh Mary, Don't You Weep" for a cry to spring to the lips. "Blues Singer" and "Bass Player" both send music soaring through the body. "Open Gate" makes one remember the many gates closed to blacks. "Mother and Child" speak poetry to the viewer. "Work," "The Harvest," and "Young Farmer" pour sweat down the brow. All of White's drawings have an emotional impact on the black viewer because they are a portrayal of the black experience. It is hard to believe that these works could not inspire some black children to poetic creations of their own.

Dance can also be an adjunct to reading. Students can pick out the central theme, the story line, the climax, and the end of a dance as one can for a story. For example, students can write stories about the soul dances from 1953 to the present, for these dances portray the black search for liberation. The "Chicken" made a mockery of the push for civil rights, for the chicken is a bird that cannot fly, a yardbird, fenced in, and defenseless. Coolness was expressed in such dances as "The Stroll," "The Madison," and "The Continental" which portrayed the disciplined and nonviolent mood of the integration move from 1955 to 1959. From 1959 to 1963 animal themes—"Horse," "Pony," "Gorilla," "Snake," "Bird," "Dog," point to the bestiality of white America. The "Flea," "Fly," and "Roach" focused on the ever-present nuisance of whites and the need to stomp that nuisence to death. "The Watusi" pointed toward an identification with the warriors of a glorious past while "The Twist" portrayed the anxieties of the period and foreshadowed the calm before the storm. Other liberation fronts were explored through the "Locomotion" which spelled farewell to black dependency on whites. The "Twine" was a search for a way around the problem and the "Hichhike" might have been a desire to reach freedom by any means. Those who wanted to use the integration route may have been depicted through the foot shuffling of the "Uncle Willie."

Since white violence in the South was the theme of 1965 and the riots came on the heels of these, the "Matador" might have shown the black reaction to charge like bulls against oppression. A turn in the struggle as characterized by the advent of Stokeley Carmichael on the scene in 1966 was shown through the "Shotgun" and the "Jerk" whose contortions revealed the black determination to face annihilation if necessary. The "Funky Broadway" repudiated the white lie about life and the "Charge" expressed the continued charge of the bull. Audaciousness and creativity poured forth in the "Boogaloo" which depicted the throwing-off of the shackles of fear.

The "African Twist" revived the continued identification with Africa as did the naturals, handshake, and the daishiki. The "Tighten Up" signified the need to close ranks against the oppressor and the "Lickin' Stick" showed disdain for the weapons of oppression. A climax was reached during the summer of 1968 in the explosiveness of the "Black Power Stomp."[48]

Any creative teacher of black literature can use these themes to inspire a literary outpouring from his students.

A FINAL WORD

Each ethnic group must define itself. Each must have his own spokesmen selected by the group from the group, for each group is captured within its own set of experiences and fashioned and molded by the rewards and punishments of society as they relate to its kind. No person outside of the contours of his group life can affirm or disaffirm his existence, can explain his feelings. Anything else is the creation of a synthetic situation.

Since a very positive self-concept does exist in black children, teachers need only to enhance this through various means. One way, as pointed out in this paper, is through the selection of reading materials and the interpretation of them in relationship to the visual arts and the performing arts, all of which can intermingle into an artistic, humanistic whole.

NOTES

1. Kenneth B. Clark and Mamie P. Clark, "Racial Identification and Preference in Negro Children," in Theodore M. Newcomb and Eugene L. Hartley, eds., *Readings in Social Psychology* (New York: Henry Holt, 1947).

2. Mary Ellen Goodman, *Race Awareness in Young Children* (Cambridge, Addison-Wesley, 1952).
3. Helen G. Trager and Marian Radke Yarrow, *They Learn What They Live* (New York: Harper, 1952).
4. Catherine Landreth and Barbara C. Johnson, "Young Children's Responses to a Picture and Inset Test Designed to Reveal Reactions to Persons of Different Skin Color," *Child Development* 24 (1953), pp. 63-80.
5. J. Kenneth Morland, "Racial Recognition by Nursery School Children in Lynchburg, Virginia," *Social Forces* 37 (1958), pp. 132-137.
6. Harold W. Stevenson and Edward C. Stewart, "A Developmental Study of Racial Awareness in Young Children," *Child Development* 29 (September 1958), pp. 399-409.
7. Richard G. Larson et al., "Kindergarten Racism: A Projective Assessment" Unpublished report, University of Wisconsin, Milwaukee, 1966.
8. Peter John Georgeoff, *The Elementary Curriculum As a Factor in Racial Understanding* (Lafayette, Ind.: Purdue, December 1967).
9. Rodney W. Roth, "The Effects of 'Black Studies' on Negro Fifth Grade Students," *The Journal of Negro Education* 38 (Fall 1969), pp. 435-439.
10. Benjamin J. Hodgkins and Robert C. Stakenas, "A Study of Self-Concepts of Negro and White Youth in Segregated Environments," *The Journal of Negro Education* 38 (1969).
11. William H. Grier and Price M. Cobbs, *Black Rage* (New York: Basic Books, 1968).
12. Roger D. Abrahams, *Deep Down in the Jungle* (Hatboro, Pa.: Folklore Associates, 1964); Elliot Liebow, *Tally's Corner: A Study of Negro Streetcorner Men* (Boston: Little, Brown, 1967).
13. Ralph Ellison, *Shadow and Act* (New York: Random House, 1953), p. 114.
14. Mary Ellen Goodman, "Evidence Concerning the Genesis of Interracial Attitudes," *American Anthropologist* 48 (1946), pp. 624-630; Kenneth B. Clark and Mamie P. Clark, "Racial Identification and Preference in Negro Children," in Theodore M. Newcomb and Eugene L. Hartley, eds., *Readings in Social Psychology* (New York: Holt, 1947); Mary Ellen Goodman, *Race Awareness in*

Young Children (Cambridge, Mass.: Addison-Wesley, 1952); Helen G. Trager and Marian Kadke Yarrow, *They Learn What They Live* (New York: Harper, 1952); Catherine Landreth and Barbara C. Johnson, "Young Children's Responses to a Picture and Inset Test Designed to Reveal Reactions to Persons of Different Skin Color," *Child Development* 24 (1953), pp. 63-80; J. Kenneth Morland, "Racial Recognition by Nursery School Children in Lynchburg, Virginia," *Social Forces* 37 (1958), pp. 132-137; Harold W. Stevenson and Edward C. Stewart, "A Developmental Study of Racial Awareness in Young Children," *Child Development* 29 (September 1958), pp. 399-409; J. Kenneth Morland, "Racial Self-Identification: A Study of Nursery School Children," *The American Catholic Sociological Review* 24 (Fall 1963), pp. 231-242; J. Kenneth Morland, "A Comparison of Race Awareness in Northern and Southern Children," *American Journal of Orthopsychiatry* 36 (January 1966), pp. 22-31.

15. Ossie Davis, "The English Language is My Enemy," *IRCD Bulletin* 5 (Summer 1969), pp. 13-15.
16. Simon Podair, "How Bigotry Builds Through Literature," *Negro Digest* 16 (March 1967), pp. 38-43.
17. Margaret Burroughs, *What Shall I Tell My Children Who Are Black?* (Chicago: M.A.A.H. Press, 1968) p. 8
18. Joseph Pentecoste, "The New Black Television—A White Strategy: A Commentary," *Inner City Issues* 1 (October 1969), p. 3.
19. Ibid., p. 5.
20. Ibid., pp. 10-11
21. Cecil M. Brown, "What Lies Ahead for Black Americans?" *Negro Digest* 19 (November 1969), p. 35.
22. Orde Coombs, "Books Noted," *Negro Digest* 18 (November 1968), p. 74.
23. Ibid., p. 75.
24. Ulf Hannerz, "Roots of Black Manhood," *Trans-action* (October 1969), p. 13.
25. Abram Kardiner and Lionel Ovesey, "Psychodynamic Inventory of the Negro Personality," in John Williams, ed., *Beyond the Angry Black* (New York: Cooper Square, 1966), p. 98.
26. Charles Keil, *Urban Blues* (Chicago: The University of Chicago, 1967), p. 27.

27. From a letter from Bernard Pettis to the author dated February 26, 1970.
28. Ibid.
29. Gerry Tyler, "Panther Party Opens Small Free Medical Clinic on Chicago West Side," *Muhammad Speaks* (February 27, 1970), p. 27.
30. Toni Anthony, "Breakfast Plan," *Chicago Daily Defender* (February 12, 1970), p. 8.
31. Hannerz, "Roots of Black Manhood," pp. 16-21.
32. William Keorapetse Kgositsile, "Young Black Poets–Paths to the Future," *Negro Digest* 17 (September/October 1968), p. 41.
33. Ibid., p. 46.
34. Don L. Lee, "Black Poetry," *Negro Digest* 17 (September/October 1968), p. 29.
35. Ibid., p. 32.
36. John A. Williams, "What Lies Ahead for Black Americans?" *Negro Digest* 19 (November 1969), p. 5.
37. Don L. Lee, "Directions for Black Writers," *The Black Scholar* 1, no. 2 (December 1969), pp. 54-55.
38. Carolyn Gerald, "What Lies Ahead for Black Americans?" *Negro Digest* 19 (November 1969), p. 25.
39. Dudley Randall, *Black Poetry: A Supplement to Anthologies Which Exclude Black Poets* (Detroit: Broadside Press, 1969).
40. Ahmed Alhamisi and Harun Kofi Wangara, *Black Arts: An Anthology of Black Creations* (Detroit: Black Arts, 1969).
41. Ed Bullins, ed., *New Plays from the Black Theatre* (New York: Bantam Books, 1969).
42. Dudley Randall, "The Broadside Voices," tapes by Knight, Randall, Arnez and Murphy, Sanchez, Eckels, Marvin X, Stephany, Lee, and Kgositsile (12651 Old Mill Place, Detroit, Michigan 48239).
43. Dudley Randall, "Broadsides," op. cit.
44. John H. Clarke, assoc. ed., *Freedomways,* 799 Broadway, Suite 544, New York, New York 10003; Daniel H. Watts, ed., *Liberator, Inc.* 244 East 46th Street, New York, New York 10017; *The Black Arts Magazine,* Concept East Publishing Company, 401 East Adams Street, Detroit, Michigan 48226; Edward Spriggs, ed., *Black Dialogue Magazine,* P.O. Box 1019, New York, New York 10027; Ed

Bullins, ed., *Black Theatre: A Periodical of the Black Theatre Movement,* Room 103, 200 West 135th Street, New York, New York 10030; Hoyt Fuller, ed., *Black World,* Johnson Publishing Company, 1820 South Michigan Avenue, Chicago, Illinois 60016; Eugene Perkins, man. ed., *Black Expressions,* 7512 South Cottage Grove, Chicago, Illinois 60619.

45. Dudley Randall, op. cit.
46. Cherry A. Banks, "Teaching The Black Experience With Popular Music," (Seattle: University of Washington, 1971), unpublished paper, p. 6.
47. *Images of Dignity:* The Drawings of Charles White (The Ward Richie Press, 1967).
48. *Rapsodi in Black: A Message in Soul Dances from 1953-1968,* Class Project, "Culture of Poverty," (Chicago: Center for Inner City Studies, Northeastern Illinois State College, vol. 1, August 10, 1969).

6

Institutional Racism: The Crucible of Black Identity

James A. Goodman

The individual's entry into the accepted roles and statuses of his society is highly complex. Given the nature of American culture, it becomes a task of immense magnitude to chart the specific steps of that journey. With respect to blacks, the social definition of race is a further complicating factor in any effort to unravel the process of identity formation.

Historically, blacks have been included in the general conceptualizations concerning identity development. However, there is growing awareness that the nature of the social patterns in our society have led to differing outcomes with reference to patterns of black and white identity. Specifically, how does discrimination on the individual or group level affect black self-identity? In this context, discrimination should be viewed more as a condition of being acted upon or attacked in the psychosocial sense rather than as ignored or neglected. Being black has many implications for the development of self-perceptions that are not consistently reflected in lower class membership.

Black people have to contend with the normal developmental tasks in addition to the survival factors associated with the fact of inheriting an inferior class status. Segregated housing, schools, and other facilities continually suggest a difference—an unacceptable one of inferiority. Recently, as an explanation for the behavior exhibited by whites towards blacks, the phrase "white racism" has become part of the daily language of a significant portion of the American public. These words join a long list of other shorthand phrases for a very comprehensive set of dysfunctional behaviors.

Racism is any individual or collective act which denies blacks access to positive identity factors in American society. However, it is of limited value to bandy this concept about in an unfocused manner. It would appear to be more useful to relate white racism to specific undesirable societal outcomes in order to provide an analytic as well as corrective frame of reference.

The focus of this discussion will be on the effects of racism in relation to the development of black identity. Black identity as an aspect of human socialization has many dimensions. We propose to look at some of the individual and group dimensions of identity as reflected in the black experience. We are particularly concerned with the experiences black people have within the educational systems of the nation.

An individual's notion of who he is contributes significantly to the development of his response pattern to the institutions of society. Further, the extent to which his notion of self is confirmed or rejected by others will be crucial to his vision of self. Personal identity, which is rooted in the processes of socialization, represents a person's search for relatedness to other individuals, groups, institutions, and practices which are sanctified in a particular cultural context. The manner in which the individual develops his personal identity statement depends both on the conditions for self-realization in the environment and the ability of the individual to perceive these conditions in objective terms. Depending upon the outcome, this may tend to enhance or impede the biological and psychological push from within that involves man's search for identity. When this inner thrust is counteracted, inner conflict may be seen as a function of the degree of awareness that the individual has. If limited alternatives are provided in the environment this conflict is likely to be expressed in "self" directed or "other" directed hostility. In order for an individual to feel a positive sense of selfhood, he must come to believe that the society in which he lives places value on his being.

SELF-IDENTITY

Some theories

Definitions and theories abound as to what constitutes self-identity. As one begins to carefully scrutinize the definitions in this area, the concept seems to become less specific. Klapp puts it this way:

> *Strictly, it includes all things a person may legitimately and reliably say about himself—his status, his name, his personality, his past life. But if his social context is unreliable, it follows that he cannot say anything legitimately and reliably about himself. His statements of identity have no more reliability than a currency which depends upon the willingness of people to recognize and accept it. We feel that we can count on our identity not only because of habit, but because we can count on people responding to it.*[1]

Erikson approaches the subject matter of identity by letting the term *identity* speak for itself in a number of connotations. He refers to a conscious sense of individual identity, to an unconscious striving for a continuity of personal character, to a criterion for the silent doings of ego synthesis, and as a maintenance of inner solidarity with a group's ideals and identity.[2] He further evaluates the definition of identity as being a psychoanalytic term as well as a psychosocial one. A distinction is then made between the self and the ego aspects of identity formation. Erikson maintains that identity covers more than what has been called the self—it includes the ego's synthesizing functions which are concerned with the "genetic continuity" of self representation.[3] Ego functions of cognition, integration, defenses, and execution perpetuate and maintain one's "idea" of oneself.

Definitions provided by Sarnoff, McCandless, and Secord illustrate the self aspect of identity in their discussion of self-concept. Self-concept is "the idea one has of oneself."[4] Self-concept involves three components of attitudes toward oneself–the cognitive, affective, and behavioral.[5] The self-concept may be thought of as "a set of expectancies, plus evaluations of the areas or behaviors with reference to which these expectancies are held."[6] Role categories are seen to aid the stability of an individual's self and behavior. Perlman suggests that problems of identity are related to some unmanageable or insufferable role difficulty.[7] The affective component of attitudes toward self and evaluation in terms of role expectancies ties in with Freud's references to self-esteem and to the ego's attitudes toward the self.[8]

The concept of identification is a crucial one in the establishment of self-identity. The developing individual incorporates and imitates the behaviors, beliefs, and values of those around him through this process. Parents serve as early models; siblings, peer groups, teachers, friends, and the larger society provide later examples. The individual's final identity is, however, more than any single input into his personality structure. The unity of self reflects the unique individual interpretation within the context of the social process.[9, 10]

Thus, self-identity in this essay will include both self and ego aspects, an idea of oneself that is perpetuated by ego functions. "Self-identity emerges from all those experiences in which a sense of temporary self-diffusion was successfully contained by a renewed and ever more realistic self-definition and social recognition.[11]

Personal and ego identity as described by Erikson are aspects of this self-identity:

> *. . . the immediate perception of one's self-sameness and continuity in time; the simultaneous perception of the fact that others recognize one's sameness and continuity. . . . the awareness of the fact that there is selfsameness and continuity to the ego's synthesizing methods and that these methods are effective in safeguarding the sameness and continuity of one's meaning for others.*[12]

In sum, self-identity is the relationship of oneself to oneself, to others, and to social institutions. It implies a continuity and sameness within the person (and perception that ego forces are effective in maintaining this) and the sharing of some essential character with others.

Not only does the individual identify with others in society, society also identifies the individual. The maturing individual soon realizes that society evaluates him, in large measure, in terms of his group identifications. Cooley posits the notion of the self as a looking glass with three related dimensions of one's self-concept: "the imagination of our appearance to the other person; the imagination of his judgment of that appearance; and some sort of self-feeling, such as pride or mortification."[13]

Self-identity as viewed by Freud and Erikson evolves through successive developmental stages. Conscious, preconscious, and unconscious factors influencing individual growth, experiences in early life that are basic to the individual's later adjustment, social learning in relation to others especially the family, ego capacities of cognition, defenses, integration, and action are all part of this developmental process that affects individual identity.

Every individual is thought to proceed through Freud's stages of psychosexual development (oral, anal, phallic, genital), either successfully or unsuccessfully. Erikson's "primary concern with the continuity of experience necessitates a shift to the function of the ego"[14] beyond these stages of Freud. Erikson rebuilds these phases so that "they lose many of their biosexual implications.[15] He stresses a wider social setting, and he focuses more upon healthy development. Erikson, then, describes eight developmental tasks that should be mastered successfully by the healthy individual. His fifth stage—achieving a sense of identity—when mastered, enables the individual to face the challenges of the adult world. His sense of identity begins to develop at birth and continues until death, but the identity crisis is in adolescence. Successful mastery of every other phase of development influences self-identity. A sense of basic trust, of autonomy, of initiative, of industry proceed that of identity; and a sense of intimacy, generativity, and integrity follow. Whatever influences successful or maladaptive solutions to each crisis of development—and specifically in regards to the obtainment of a sense of identity—we know that the individual and his environment interact in all stages of the individual's development.

Certain factors influencing adaptive patterns in an individual's development are applicable to this particular subject. Forces within the individual respond to outside influences–especially those of other persons. As an infant the child's contact is generally with his family, primarily the mother. The child's experience with his environment will result in his perception of the world as basically comforting and secure or as basically hostile and fearful. What the child will "dare" in terms of new experiences will be much molded by his earlier experience. His own worth and ability is reflected by reactions of others toward him. While many authors indicate that the foundation of character is established by five years of age and that disordered relationships early in life may leave a nearly ineradicable scar, this writer agrees with Allport in his view that the child who "enjoys a normal affiliative groundwork" has established foundations of character only in the sense that he is then "free to become."[16] Erikson also has a more optimistic view of human growth as he emphasizes task resolutions throughout the life of the individual.

In sum, then, the process of identity formation in Erikson's formulation emerges as an evolving configuration:

> *. . . it is a configuration gradually integrating constitutional givens, idiosyncratic libidinal needs, favored capacities, significant identifications, effective defenses, successful sublimations, and consistent roles.*[17]

The interplay between the individual and his family as well as the larger society should permit the individual to establish a self-identity recognized mutually by that individual and his society.

Obviously, individuals are rewarded differentially on the basis of their memberships in various groups. Cultural patterns in the United States, for example, have long provided a basis for blacks to be evaluated as being different from, and outside of, the norms of society. Identity formation for blacks, with respect to the larger society, is rooted in a societal statement of scorn and disparagement. The black individual is taught that his selfhood must articulate a different imposed notion of who he is. Therefore, the black individual who identifies with the predominant cultural values risks accepting and ritualizing the very behaviors which place limits on what he may become.

In this country the black individual is given a consistent

image of who he is by white society; black skin means that he if inferior and not quite human. The black individual comes to evaluate himself in the context of his significant holding groups with the knowledge that racial and ethnic groupings constitute the primary limits of his identity. The influence of membership in the black group, for instance, supersedes the consequences of memberships in all other groups.

Several other theories of socialization provide additional insight into the complex identity-formation processes. The contributions of learning theory are of particular usefulness to this discussion.

LEARNING THEORY AND IDENTITY DEVELOPMENT

Learning theory in relation to the development of self-identity is limited here primarily to the formulations of Robert Sears; his formulations have utilized analytic concepts and complement theories of psychoanalysis and ego psychology. "From the learning point of view, the self-concept is the apex—the culmination of all the social and personal experiences the child has had.[18] Conditioning and instrumental learning, primary and secondary generalization, reward and punishment, motives and drives, expectancies and probabilities, and conflicts all are involved in establishing identity. The satisfactions of fears experienced by the individual (throughout his total life experience—but most importantly in current reality) develops the framework for individual identity.

Environment shapes behavior; stress is placed upon parental child-rearing practices. Behavior is seen as self-motivated by its tension-reduction effect: every unit of behavior preceding a goal achieves a reinforcement potential. Primary drives are only instrumental for the beginning of behavior in a social world—socially learned (secondary motivational) systems eventually motivate all behavior. Since all behavior represents reinforced actions, development is viewed as a training process. Parents are seen as the most important reinforcing agents; Sears maintains that every "parent could do better if he knew better."[19] Parents must then have access to reinforcement procedures that would contribute to healthy development. Sears speaks also of the effect of learning beyond the family upon the socializing of the child. The individual then within a learning theory framework, would gain self-identity through reinforcement of behavior by others.

In addition, Sears mentions that identification influences behavior, which starts with the mother-child relationship and extends to others as the child becomes older. Identification rests neither upon trial-and-error nor child-rearing efforts—it evolves from the child's own role playing. The child selects available others as models and imitates them. Reinforcement comes through recognition received from others for imitating and personal satisfaction in seeing the actions of others in one's own behavior.

Self-identity can then arise from satisfactions in perceiving of oneself, as behaving as another, and from responding to the expectations of behavior as conveyed by others.

Although Sears focuses much upon behavior, effective identity involves cognitive, affective, and action components. Self-identity also implies a continuity of identity as well as an awareness of (or idea of) identity. Therefore, Erikson's formulations including both ego and self aspects appears to this writer to integrate learning and analytic theories to successfully designate the process and definition of self-identity.

Unfortunately, the process of incorporating the societal statements about self and self-potential begin very early. Allport maintains that a child is capable of a sense of ethnic group identification as early as the age of five.[20] In a fairly extensive study of 253 black children between the ages of three and seven, Clark and Clark maintain that the period between four and five years of age may be the critical period in the development and outlining of racial attitudes towards oneself and others. These researchers posit: "At these ages these subjects appear to be reacting more uncritically in a definite structuring of attitudes which conforms with the accepted racial values and mores of the larger environment."[21]

DEVELOPMENT OF BLACK IDENTITY

The black individual's concern with self-identity, unlike that of his white counterpart, is supposedly emotionally laden in all of its aspects. As Proshansky and Newton put it:

> *. . . the young child acquires value-laden racial labels and fragments of popular stereotypes to describe his own and*

> *other racial and ethnic groups. Both Negro and white children learn to associate Negro with "dirty," "bad," and "ugly," and white with "clean," "nice," and "good." For the Negro child, these emotionally charged descriptions and judgments operate to establish the white group as vastly superior to his own racial group.*[22]

Identification with the majority culture, it would seem, is not a method of survival for black people. At best it suggests accommodation to the status quo. It would appear that identification with the aggressor is a threat to healthy black self-identity.

A balanced view of self is the positive outcome of having access to a wide range of roles. The individual with this type of identity knows himself to be a full-fledged member of the communal culture. For blacks, however, the limited range of available roles affects the development of positive self-identity in direct relationship to the individual's reliance upon those cultural symbols which inevitably place him at odds with the reality associated with his blackness. In the Eriksonian model, identity versus identity diffusion is the choice to make when childhood proper comes to an end and youth begins. Growing and developing young people are then primarily concerned with attmepts at consolidating their social roles. They are sometimes morbidly, often curiously preoccupied with what they appear to be in the eyes of others as compared with what they feel they are.[23]

It is within this struggle for self-acceptance that the black individual fares badly; the eyes of whites reflect back a nonexistent or limited sense of cultural continuity. The "old country" is stripped of all validity in terms that are meaningful to him. He has therefore found it convenient to define himself in terms that are alien to his personal frame of reference; his selfhood is related to a paradoxical statement: in order for black identity to have meaning it must exist only as a reaction to white identity.

The black father, the black mother—the sacred hosts of all black progeny—give mute testimony to the progression of genius shunted into meaningless and repetitive acts of mediocrity. These acts, these obscenities, give rise to the general cultural statement that blacks are completely incapable of self-realization outside the context of the white-oriented frame of reference. This argument suggests that the peculiarities of the black subculture constitute a variant of the general societal pattern of identity development. Logically, however, one would expect that the process of black socialization would produce patterns of identity substantially different from those associated with whites.

Although the data concerning changing patterns of black identity are not fully available, it appears that blacks have begun to reject imposed definitions of self. Individual black identity thereby becomes a potent force in black group consciousness. Individuals begin to relate their selfhood to the blackness variable as a statement of group unity and cohesiveness. Arnold Rose defines group identification as a conscious recognition that one is a member of a group with a positive evaluation accorded such membership.[24] This identification or sense of unity in a minority group is created by pressure from outside the group.

In the instance of blacks, the myriad of negative symbols conjured up by the white society served as the locus for developing a different statement of selfhood. The concept that "black is beautiful" is merely the overt manifestation of the deeper process of trying on new, and previously denied, role patterns. The research by the Clarks makes it very clear that black youngsters in the past did not consider black as being beautiful. This research was conducted prior to the beginning of what is commonly referred to as the black revolution. The period of the sixties has seen an explosion of black pride which will have significance for the future development of black identity.

The future development of notions of selfhood by blacks will be more than a function of the black is beautiful phenomenon. However, as indicated earlier, identity is an outcome of an interactional process between the individual and society. The social process of confirming the black individual in society is enhanced in the primary social institutions of the society. Billingsley states:

> *Racism is deeply imbedded within the institutional fabric of American society. All the major institutions including the political, economic, educational, social, and others have systematically excluded the Negro people in varying degrees from equal participation in the rewards of these institutions. None of them works as effectively in meeting the needs of Negro families as they do white families. The keys to the enhancement of Negro family and community life are therefore institutional keys.*[25]

Obviously, the quality and function of social institutions can have impact upon the nature of black selfhood. If these outcomes are compatible with black societal prerequisites for survival as a viable cultural entity, there is no argument. It is when the institutional arrangements are such that black people have limited access to them, or upon relating to them find that they impede the development of positive black identity, that black social concern develops. This concern has to be translated into acts designed to alter the negative impact these institutions have on the growth and development of black identity.

It is quite clear that the institutional advantages enjoyed by whites create the general cultural assumption that blacks constitute a lower socioeconomic group which is monolithic in character. This factor probably contributes most significantly to the white community's attitudes that it is superior to the black community in all areas. This form of racism, when translated into the historical quality of contacts between blacks and whites, also serves to reinforce the black individual's notions of personal and community inferiority.

Although it is true that blacks are more likely to be afflicted with the ills of poverty and low socioeconomic status than whites, it is equally true that blacks learn to cope in highly unique ways. Moreover, the effect of black people's perception of their class position as this affects identity has not been made clear. The basic question is: how can a wider range of coping strategies be provided while racist practices solidify the barriers against entry into the institutional life of the country? This question becomes the focal point for various strategies to be posed in the process of liberating blacks from the vortex of an identity rooted in the American way of life. For example, should blacks try to change the total society? Perhaps the focus should be upon key institutional practices. Because blacks are a diverse people, their responses to institutional racism will likely reflect this diversity.

Himes illustrates the nature of the issue for some:

> *In the case of lower-class Negroes, the significant institutional preconditions include, among others, color segregation, material discrimination, inferior or collateral social status, disparaging social evaluation, chronic social frustrations, and a substantially distinct subculture. From socialization from such preconditions, the individual emerges as a*

functioning member of his social world. Certain dimensions of the functional adjustment to his effective social world, however, constitute cultural deprivation in terms of the standards and demands of the larger world from which he is more or less excluded.[26]

EDUCATIONAL INSTITUTIONS AND BLACK SELF-CONCEPT

The major institution of socialization in this society in addition to the family is education. In many ways the family has given responsibility for the supervision of the child's development to the school. It is clear, therefore, that at least the school and the home share responsibility in varying degrees for the socialization of the child. As McNeil views it:

> *The terms "education" and "socialization" should be considered synonymous in our society, for education is the primary means of socializing all children after they reach the age of five. Our children now spend the bulk of their time in groups of about 30 strange peers dominated by a professional teacher in a building specially erected for that purpose. This educational system socializes children by teaching them the knowledge and intellectual skills essential to full participation in society as well as the mores and habits of its members.*[27]

Education must, therefore, reflect the nature of the total society if it is to prepare all individuals to live in their effective social worlds. This formulation implies an egalitarian approach within the educational delivery system. However, all segments of the population in the United States have never received the same quality of education. It is only recently that this inequality has been viewed as problematic. For a long time, Americans were content with the doctrine of "separate but equal" educational facilities for that part of its population considered inferior—the blacks. Because blacks were considered inferior, and therefore unteachable, it was thought that admitting them to white schools would lower the quality of education received by the whites. As a consequence of this type of thinking, separate schools for blacks were established; there was, however, limited concern for the quality of these schools. Since the school system in this country is undeniably an arm of the government, political pressure is the reason for each school being as it was and is now—good or bad.[28]

In 1954 the Supreme Court decided that separate but equal educational facilities were unconstitutional, and ordered the schools to desegregate. This decision reflected a growing awareness that educational goals for blacks were not being achieved in the facilities allocated to them. In a majority of cases their potential has not been realized because of environmental factors, such as family and neighborhood deprivation, differences in cultural tradition, and economic impoverishment.[29]

Educators have kept blacks from realizing their potential by assuming that they cannot expect much of them and by treating them condescendingly. The children find it natural and automatic to accept the school's structural inadequacies and to incorporate them, as it were, into their notions of self. Many of these youngsters actually begin to view themselves as biologically inferior to whites.[30]

More people gradually have come to accept the *ideal* of educational equality for all and to realize that our school system must compensate with a saturation of services that will rescue the culturally different (black) youngsters emotionally, provide them with direction, and above all, inculcate skills enabling them to function successfully both academically and vocationally.[31] But de facto segregation continues to exist. Because segregated schools often reflect patterns of segregated housing, many people feel that the schools cannot do anything unless whites and blacks are geographically integrated. Others suggest various bussing plans as a means of achieving quality education for all youngsters.

In the North, much of the effort of the civil rights movement has been directed toward the elimination of de facto school segregation through some form of modification in the traditional patterns of school districting. Implicit in the activities designed to alter these traditional patterns is the assumption that integrated education produces quality education for students without regard to their racial backgrounds. As Young puts it:

> *. . . with integration Negroes will benefit from a better educational system, better materials, facilities, and teachers. But as I see it, there are precious benefits for both whites and Negroes. I do not believe education can be absolutely first rate without integration. To the extent that people say that integration of the schools has no relationship to the quality of education, they are, in fact, saying that separate but equal education is valid.*[32]

A counter argument, of course, is that schools in all areas of the community should be brought to a single standard of excellence. Youngsters would then receive quality education without respect to their place of residence. On this point, however, Clark argues that, "the goals of integration and quality education must be sought together; they are interdependent. One is not possible without the other."[33] The validity of either argument has to be evaluated in terms of the functions of education. The mere presence of blacks and whites in the same classroom is not a guarantee that change will occur in the educational content provided. Stated differently, integration cannot of itself solve the problems associated with racial isolation. Integrated schools can, however, provide the opportunity for black and white youngsters alike to develop equal status contacts which alone can diminish the effectiveness of racially created antagonisms and distance producing behaviors.

From the vantage point of white society, the primary function of education is the maintenance of white culture. Ostrom makes the point in this manner:

> *Man's capacity to learn, to organize learning in symbolic forms, to communicate this learning as knowledge to other members of the species and to act on the basis of learning or knowledge is the source of all cultural phenomena. . . . Any culture and the civilization based upon that culture must depend upon the ability of the civilization to articulate and transmit its learning as semiautonomous, cognitive*

> *systems. These represent the accumulated knowledge in every field of inquiry and comprise the subject matter in all education. This is what we mean when we speak of the school's responsibility in transmitting a cultural heritage.*[34]

Some underlying values and assumptions

If there is ever going to be more general agreement about what is to happen when youngsters come together in designated educational space, the teaching content and methodology must not be based on the assumption that the white middle-class subculture is better than the black subculture. At best, each subculture is functional to the needs of the respective groups. The white-dominated educational system must recognize the particularized needs of the black student that evolve from the black subculture which previously has been ignored in planning educational activities. Ornstein[35] has suggested that the beginning of a solution is to do away with the strictly middle-class standards of the school. He feels that these standards do more harm than good in lower class schools.

The educational system must accept the disadvantaged on their own terms and work to achieve its goals by serving as an ego-supporting, meaningful institution which encourages diversity. This would enable the disadvantaged to become their best possible selves by utilizing their culture, not by trying to change it.

Basically, what is indicated for the educational system is the provision of a learning opportunity for black youngsters to gain a positive social identity. Social identity, in this context, refers to the expression of selfhood which is an external manifestation of the black individual's perception of who he is—of allowable social roles. In discussing this aspect of self we must be mindful that it does not contain the total statement of the individual's integrity. There are, as we suggested earlier, other aspects of identity. The holistic quality of personal identity is of great significance in relation to blacks because of the negative cultural statements regarding the black presence in the society. Society, which is generally benign toward people, deprives blacks of man's most fundamental right; that of self-determination through creative activity. This is only possible if the individual is accorded dignity so that he is free to make choices over a wide expanse of opportunities.

Schools, as primary agents of socialization, must help black youngsters attain a positive view of self at the internal level of identity. This inner quality of identity is the outcome of the responses to the cultural environment. This aspect of personal identity is highly critical; black inner identity is a result of the many responses to the cultural statement of black group identity.

Apart from its broad function of transmitting culture in a formal manner, education has a role in influencing the course of social change; determining the content of culture for future generations of students is of equal importance. There has to be greater recognition that the school is very much a part of society and must be criticized for not concerning itself with being implicated in the major social problems of our society. As a part of society, the school has been involved in discriminatory practices. It has perpetuated the racism which fostered segregation in all aspects of American life, including housing. When the courts order the schools to remedy the consequences of housing patterns, they are actually directing schools to make amends for past acts of commission and omission which have culminated in the ghettoization of American blacks.

Black schools are academically inferior because they reflect the cumulative inferiority of segregated education and the inevitable problems of a racist, segregated society which imposes upon lower-status individuals a debilitating, humanly destructive form of public education, both in the South and the North. Black schools are inferior because our society persists in not finding the commitment or the resources to provide high quality education for powerless black youngsters—because associated with this rejection, exclusion, and the dehumanizing aspects of racism, is the inevitable lowering of morale in any lower-status institution. Lower-status schools present crises in self-respect, nagging and gnawing feelings of inferiority, or deep and disturbing questions related to self-hatred. These types of schools add to the black youngster's rejection of self. Proshansky and Newton report empirical evidence that shows this tendency toward self-rejection:

> *. . . . Given a choice, a majority of both Negro and white children tend to choose a white doll in preference to a Negro one (Clark and Clark, 1947; Stevenson and Steward, 1958; Radke and Trager, 1950; Goodman, 1952; Morland,*

> *1962; Landreth and Johnson, 1953). In a more recent study of 407 young children, Morland (1962) found that 60 percent of the Negro children, but only 10 percent of the white children, preferred to play with children of the other race; in comparison, 18 percent of the Negro children and 72 percent of the white children preferred playmates of their own race.*[36]

These data and the attendant conclusions are bolstered by the fact that large numbers of Negroes use products which are designed to make them into imitation whites. In nearly all popular black magazines one encounters advertisements for skin creams, bleaches, pomades, and other compounds designed to transform blacks into an acceptable white facsimile. This argument is not designed to suggest that the school is fully responsible for this set of circumstances. The school, however, continues to create the illusion that only middle-class white norms of behavior are acceptable. Therefore, the black child is forced to abandon his identity if he is to be rewarded by society. The white individual does not go unscarred in this process. Clark gives the following reasons why he also considers white schools inferior:

> 1. Education has become ruthlessly competitive and anxiety-producing, in which the possibility of empathy and the use of superior intelligence as a social trust are excluded from the educational process.
>
> 2. They have facilitated the reduction of the educational process to a level of content retention required for the necessary scores on college boards.
>
> 3. They have permitted elementary and secondary schools to become contaminated by and organized in terms of the educationally irrelevant factors of race and economic status.
>
> 4. They have watched in silence the creeping blight of our cities and the spawning of Negro ghettos.
>
> 5. They have remained detached and non-relevant to this major domestic issue of our time.[37]

It is clear that education performs a major socializing function in society, and that this is one of its primary functions. We may look to the schools to teach children reading, writing, and arithmetic, but we must admit that we also look to the schools for the inculcation of a particular civic culture. The schools are expected to instill a set of normative values which support, not challenge, the existing societal values.[38]

If, as we posit in our thesis, the normative values which the schools are expected to instill are middle-class and white-oriented, then schools socialize black youngsters into the existing societal values which are rooted in racism. The fact that the American educational delivery system has been designed for the youngsters of the white middle-class sector of America forces the black youngster to see himself as unimportant and therefore excluded. The black experience as part of the cultural statement emanating from the pluralistic nature of American society is conspicuous by its absence from the educational process. Black students, from kindergarten to graduate school are expected to engage psychologically, culturally, and socially in an endeavor which is ostensibly designed for the purpose of broadening their horizons and providing them proper knowledge to better relate to the world. In truth, educational enterprises are unaccepting and essentially rejecting of the social and cultural backgrounds of black people.

Class and family considerations

Schools must serve as a bridge between the lower-class black family and the demands of potentially hostile white institutions. The family is consistently ascribed by most writers to be a major source of self-identification. The question arises: how does the identification process in lower-class black families differ from the largely white middle-class family?

With respect to class differences, the problem of just who is in the lower class needs clarification. For our purposes, Gans' and Miller's description of the lower-class subculture will be used.[39] There is disagreement not only about who is in the middle class, but also in regard to the values, child-rearing practices, and behavior which can rightfully be attributed to it.

Two basic views and sources of research stand out in regard to values. Writers such as W. Miller, L. Hyman, A. Davis, L. Empry, and

O. Lewis, indicate that the lower-class culture has a stable tradition with integrity of its own.[40] Others such as Rodman, Parsons, and Merton refer to the common value system in all of society.[41] Rodman talks about a lower-class value sketch and concludes that the lower class does share the values of the overall society, but by adapting to his circumstances, the lower-class person holds these values less strongly and also develops new values unique to the lower class. Recent writing has maintained that lower-class values are concerned:

1. with avoiding trouble, with toughness, smartness, excitement, fate, and autonomy
2. with lower levels of aspiration
3. with high education, but more in terms of improved income rather than satisfaction in work accomplished
4. with similar desires for leisure time and security
5. with adapting values to deprived circumstances[42]

Child-rearing practices of the lower-class also continue to be questioned. However, lower-class families are viewed as:

1. more involved in punishment during toilet training
2. expecting lower educational achievements of their children
3. more concerned with respectable behavior in children—obedience, neatness, and cleanliness—rather than developing self-reliance and independence
4. attuned primarily to maintenance functions
5. rigid and oriented to discipline in terms of roles and parent-child relations[43]

Lower-class behavior in contrast to middle-class behavior is said to be:

1. composed of more hostility, tension, and aggression
2. segregated as to sexes

3. more intolerant and authoritarian
4. given to resignation and fatalism
5. more prone to action
6. present rather than future-oriented
7. productive of more illegitimate birth—this situation is not condoned but is more prevalent in lower-class families
8. in a state of alienation—having feelings of powerlessness, meaninglessness, anomia, and isolation[44]

Family structure in the lower class is described as:

1. primarily female-based
2. having patterns of serial marriages
3. role of the male is marginal
4. emphasizing peer group relationships—grouped by age and sex
5. higher evaluation of female than male role
6. differing from both the working class, with its family-circle orientation (both parents and relatives) and the middle class, with participation in family and the larger society, by its (lower class) focus upon the mother and female relatives[45]

Although lower-class families often have a good deal of concern for one another, an ability to share, and less insistent valuation on conformity, achievement, and "success"; they often lack opportunities to expand their experiences. This lack of opportunity extends into the educational context. Levine sees the ineffectiveness of the school in teaching lower-class (black) youth as deriving not so much from its commitment to values which are foreign to the youth, as from its failure to provide the type of environment which would reinforce their groping and half-hearted attempts to live up to these goals.[46]

Clearly, the black individual's behavior in families is closely structured by adaptation to deprivation and lack of opportunity. Self-evaluation would obviously demonstrate lower-class categories. Lack of economic security, improper nutrition, poor housing, and less adequate clothing often produce and reflect a lowered self-image in comparison to the larger society.

Family feeling of powerlessness, fatalism, and fear of planning can limit individual goal-seeking. The middle-class child can more easily draw upon models from the school, from occupations other than the relatively isolated lower-class black child.

In relation to Erikson's formulations, one finds the black child at a disadvantage because the economic insecurity of his home often does not enable him to develop an adequate sense of societal trust. He is also less likely to develop a sense of autonomy and initiative because lower-class black families often have ambivalent attitudes toward child independence and fewer opportunities. More emphasis is placed on luck within the black family because a world controlled by fate that produces failure is often perceived.

Difficulties experienced in mastering each of these earlier stages also make self-identity a more complex task. Obviously, many lower-class black families can and do succeed in aiding the child to successfully master these phases of development. Ausubel has emphasized that "the consequences of membership in a stigmatized racial group can be cushioned in part by a foundation of intrinsic self-esteem established in the home."[47] However, lower-class black families (in common with all blacks) that have to overcome deprivations of poverty plus those of racism, do have the momentous task of aiding the child to establish a positive and healthy self-identity.

Because the school is such an integral part of society, because it places people in the reward system, it must develop the social as well as the individual man. The current concern with education of the culturally different (black) child is to make education relevant to his needs as an individual and as a social being.

NOTES

1. Orrin E. Klapp, *Collective Search for Identity* (New York: Holt, 1969), pp. 5-6.
2. Erik Erikson, "Identity and the Life Cycle," *Psychological Issues* 1, no. 15 (1959), Monograph 1, p. 102.

3. Erikson, ibid., pp. 147-149.
4. Irving Sarnoff, *Personality Dynamics and Development* (New York: Wiley, 1962), p. 142.
5. Paul Secord and Carl Backman, *Social Psychology* (New York: McGraw-Hill, 1964), p. 579.
6. Boyd R. McCandless, *Children and Adolescents* (New York: Holt, 1961), p. 174.
7. Helen Harris Perlman, "Identity Problems, Role, and Casework Treatment," *Social Service Review* 37, no.3 (September 1963).
8. Erikson, op. cit., p. 147.
9. George H. Mead, *Mind, Self, and Society* (Chicago: University of Chicago, 1934), Part 3.
10. Charles H. Cooley, *Human Nature and the Social Order* (Glencoe, Ill.: Free Press, 1956), passim.
11. Erikson, op. cit., p. 149.
12. Erikson, ibid., p. 23
13. Cooley, op. cit., p. 184.
14. Henry Maier, *Three Theories of Child Development* (New York: Harper & Row, 1965), p. 17.
15. Maier, ibid., p. 17.
16. Gordon Allport, *Becoming* (New Haven: Yale University, 1955), p. 33.
17. Erikson, op. cit., p. 116
18. McCandless, op. cit., p. 172.
19. Maier, op. cit., p. 150.
20. Gordon Allport, *The Nature of Prejudice* (Garden City: Doubleday, 1958), pp. 28-29.
21. K. B. Clark and M. P. Clark, "Racial Identification and Preferences in Negro Children," T. M. Newcomb and E. L. Hartley, eds., *Readings in Social Psychology* (New York: Holt, 1947), pp. 169-178.
22. H. Proshansky and P. Newton, "The Nature and Meaning of Negro Self-Identity," M. Deutsch et al., eds., *Social Class, Race and Psychological Development* (New York: Holt, 1968), p. 186.
23. David J. De Levita, *The Concept of Identity* (New York: Basic Books, 1965), p. 62.
24. Arnold Rose, *Sociology: The Study of Human Relations* (New York: Knopf, 1965), p. 684.
25. Andrew Billingsley, *Black Families in White America* (Englewood Cliffs, N.J.: Prentice-Hall, 1968), p. 152.

26. Joseph G. Himes, "Some Work-Related Cultural Deprivations of Lower-Class Negro Youths," Louis A. Ferman et al., eds., *Poverty in America* (Ann Arbor: University of Michigan, 1965), pp. 384-385.
27. Elton B. McNeil, *Human Socialization* (Belmont, Calif.: Brooks/Cole, 1969), p. 138.
28. Lydia Pulsipher, "The American School: A Legitimate Instrument for Social Change," *School and Society* 96 (March 30, 1968), p. 201.
29. Regina Barnes, "Higher Horizons," *Clearing House* 40 (October 1965), p. 113.
30. Jonathan Kozol, "Halls of Darkness: In the Ghetto Schools," *Harvard Educational Review* 37 (Summer 1967), p. 393.
31. Ibid., p. 393.
32. Whitney M. Young, Jr., *To Be Equal* (New York: McGraw-Hill, 1964), p. 110.
33. Kenneth B. Clark, *Dark Ghetto: Dilemmas of Social Power* (New York: Harper & Row, 1965), p. 117.
34. Vincent Ostrom, "Education and Politics," B. Henry Nelson, ed., *Social Forces Influencing American Education,* (Chicago: National Society for the Study of Education, 1961), pp. 10-12. Distributed by the University of Chicago Press.
35. Allen Ornstein, "Reaching the Disadvantaged," *School and Society* 96 (March 30, 1968), p. 215.
36. Proshansky and Newton, op. cit., p. 187.
37. Kenneth B. Clark, "Higher Education for Negroes: Challenge and Prospects," *Journal of Negro Education* 36 (Summer 1967), p. 200.
38. Charles V. Hamilton, "Education in the Black Community: An Examination of the Realities," *Freedomways* 8 (Fall 1968), p. 319.
39. Herbert J. Gans, "Subculture and Class," *Poverty in America,* Ferman et al., eds. (Ann Arbor: University of Michigan, 1965), pp. 302-311. See also: Walter Miller, as quoted in "Implications of Urban Low-Class Culture for Social Work," *Social Service Review* (September 1959), p. 230.
40. See review in Hyman Rodman, "The Lower Class Value Stretch," *Poverty in America,* op. cit., pp. 270-285.
41. Rodman, ibid., p. 270-285.

42. W. Miller, op. cit., pp. 219-236. See also: "Focal Concerns of Lower-Class Culture," in *Poverty in America,* op. cit., pp. 261-270; L. Hyman and A. Davis, in *Poverty in America,* p. 273; Lola Irelan, ed., *Low Income Life Styles* (Washington, D.C.: H.E.W., Div. of Research, 1966), pp. 6-7.
43. Irelan, Lola, op. cit., pp. 15-31. See also: R. Havighurst and A. Davis, "A Comparison of the Chicago and Harvard Studies of Social Class Differences in Child Rearing," *American Sociological Review* 20 (August 1955), pp. 438-442; E. Herzog, "Some Assumptions About the Poor," *Social Service Review* 37, no. 4 (December 1963), pp. 389-401.
44. Herzog, op. cit., pp. 389-401; Irelan, op. cit., pp. 1-26; J. Pakter, "Out of Wedlock Births in New York City," *American Journal of Public Health* 51, no. 5 (May 1961), pp. 683-697.
45. W. Miller, "Implications of Urban Low-Class Culture for Social Work," and *Poverty in America,* op. cit.; Herzog, op. cit. See also: "Is There a Breakdown of the Negro Family?" *Social Work* (February 1966), reprint from Anti-Defamation League of B'nai B'rith, New York; Gans, *Poverty in America,* op. cit., pp. 302-311.
46. Daniel Levine, "Cultural Diffraction in the Social System of the Low Income School," *School and Society* 96 (March 20, 1968), p. 206.
47. D. P. Ausubel, "Ego Development Among Segregated Negro Children," *Mental Hygiene* 42 (1958), p. 368.

7

Social Science and Education for a Black Identity

Barbara A. Sizemore

Like other excluded groups, black people in the United States of America struggle for political, economic, and cultural equality. This struggle has not yet delivered the desired fruits because blacks have internalized the dysfunctional white value system. Additionally, the set of knowledge necessary for the construction of a new value system is absent. The producers of new sets of knowledge are the scientists. The purveyors of those sets are the teachers who use the schools for dissemination and distribution. Teachers then, become the prime agents for training blacks to have the same prejudices and negative attitudes toward blacks that whites have,[1] and teachers learn their content and methods from the social scientists who create the educational models used to develop self-concept, identity, and self-image.

Recently, a great awakening occurred. It has been described as a new-found pride in blackness and the African heritage. This awakening has spurred a search for a new set of values and a lost identity[2] and has produced drives for enculturation instead of acculturation, for liberation rather than integration, and for ethnic preservation not ethnic assimilation. The problem is how to be black in a universe which denies blacks humanity. What is the black identity in such a hostile environment?

These questions indicate a need for scientific research on the hostility of the environment. Yet social science has dealt inadequately with this enigma. This paper will attempt to discuss the problems of social science research and a new theoretical framework; to define value; to review several studies which promote error; to define the concepts necessary for educating black and minority group people; and finally, to make some observations concerning that education.

SOCIAL SCIENCE RESEARCH: PROBLEMS AND THEORY

Social scientists are frequently accused of a collective bias toward western European civilization and the white race, thereby invalidating the so-called objectivity of social science research. Blacks show an unwillingness to accept the findings of these scholars and the solutions emanating therefrom.[3] Why does the black man feel that social science has failed him?

First, science has two main jobs according to Homans: discovery and explanation. Discovery is the job of stating and testing more or less general relationships between properties of nature, and explanation is the process of showing that the finding follows as a logical conclusion, as a deduction, from one or more general propositions under specified given conditions. This relationship, then, provides a theory. Repeatedly, Homans argues that social science is in trouble where explanations are concerned.[4]

Commenting that nonoperating definitions and orienting statements add to the confusion, he describes the former as nontestable variables and the latter as statements which fail to predict and explain, specifying only that something will occur but not what.[5] Value, culture, and identity are nonoperating definitions which figure largely in orienting statements. Their necessity is one of the great problems of social science. But, as Merton and Homans both indicate, such statements are necessary "approaches" even though they are not "arrivals." Homans' plea is to deliver:

> *. . . But sooner or later a science must actually stick its neck out and say something definite. If there is a change in x, what sort of change will occur in y? Don't just tell me there will be some change. Tell me what change. Stand and deliver!*[6]

Homans' advice is to reduce the study of man to the individual through the adoption of psychological propositions in a single science utilizing the determinist philosophy. Deploring the lack of honesty in social science, he begs for the admission that "our actual explanations are our actual theories."[7]

While discouraging an adoption of determinism, Yankelovich and Barrett tend to agree with the single science-of-human-nature idea.[8] Specifically interested in psychoanalysis, their criticism of research is threefold. First, the methodology and philosophy of the physical sciences are not appropriate for the study of human nature. Secondly, further pursuit of the nature-nurture controversy in its present form is useless; and third, the metatheory is not broad enough to encompass a study of man.

Pointing out the debate between Sartre and Lévi-Strauss, Yankelovich and Barrett discuss the futility of continuing this controversy as long as the premises on which it is based are invalid. Those premises are: (a) a conception of freedom which does not take human limits into account, and (b) a conception of nature which reduces its richness and complexity to a mere energy-dissipating mechanism.[9]

They propose a man-in-his-environment approach based on the doctrine of synergism, which takes epigenesis into account. They describe synergism as the process whereby every trait of personality, every meaningful human relationship, every lasting social innovation, introduces a structure that has never existed before. Through synergism, new structures appear, and all "explanations" of such structures must describe this new element.[10] Epigenesis is the process of the development of structures which grow when phylogenetic factors interact with critical individual experience at specific stages in the life cycle. They use Erikson's definition of the concept.[11]

Erik H. Erikson is well known for his concept of the epigenetic development of each individual in eight stages: (1) basic trust versus mistrust; (2) autonomy versus shame; (3) initiative versus guilt; (4) industry versus inferiority; (5) identity versus identity confusion; (6) intimacy versus isolation; (7) generativity versus stagnation; and (8) integrity versus despair. Erikson's epigenetic principle is derived from the growth of organisms and states that anything that grows has a ground plan, and that out of this ground plan the parts arise, each part having its time of special ascendancy, until all parts have arisen to form a functioning whole.[12]

Accordingly, Yankelovich and Barrett describe explanation as a special kind of description showing a tightly disciplined selectivity, presenting a case or a theory at several levels of generality at the same time, and including purpose as well as process. It would no longer be causal explanation, mimicking the physical sciences. In fact, there may be even an element of indeterminacy in human experience, and "predicting the path of a bullet in space is not the same as predicting the aftermath of a human tragedy that the bullet brings about." Also, objects, defined narrowly, have only a mechanical aspect while human experience has both a mechanical and a synergistic aspect; and, human experience requires another kind of explanation. Yankelovich and Barrett hold that "psychology as a science should be based on a theory that will not be undermined if some margin of indeterminism figures in the picture."[13]

They freely admit "that the precise details of change over a long period of time in an individual or group grow ever more difficult to predict as synergistic possibilities increase." Furthermore, the development of new structures is greatly affected by the presence or absence of specific individual experiences at the appropriate time in a given environment.[14] This frame of reference, synergism combined with epigenesis, may afford more opportunities to study the human condition of western civilization and its derivatives than present social science theories do.

Yet, there exists another problem to plague social science. The research done by Rosenthal and Jacobsen revealed the effect of the expectations of experimenters on their findings. Rosenthal and Jacobsen called this the self-fulfilling prophecy. In the reports of the experimenters, themselves, those who had been led to expect better performance from their rats viewed their animals as brighter, more pleasant, and more likeable. They reported more relaxation in their contacts with the animals and described their behavior toward them as more pleasant, friendly, enthusiastic, and less talkative. They also stated that they handled their rats more and also more gently than did the experimenters expecting poor performance.[15] Rosenthal and Jacobsen seem to show that the researcher himself is important to the outcome of the research.

Moreover, their findings seem to apply to Yankelovich and Barrett's scholarly activity; for, in spite of the fact that their new theory provides them with a view of man in his environment, takes into account evolution and culture, admits the incongruity of determinism versus freedom, aspires toward a theory of human nature, Yankelovich and Barrett both conclude that the values of the Protestant ethic are worthy because they have built a great civilization.[16] Excluded from this conclusion are the facts that the exalted virtues of the Protestant ethic–the primacy of impulse control over expressiveness; the virtue of husbanding one's resources prudently for the future; and the high value of work and calculation–produced a competitive economic model which necessitated the rape and murder of the Indian for his land and the creation of the most brutal form of slavery known to man in order to capitalize on it. So the experimenter's value system determines his findings, and social science must design a control for this effect.

THE PROBLEM OF VALUES

Wheelis suggests a conceptualization of value which offers the opportunity to examine alternate sets of values at the same time. He explains that there are institutional values and instrumental values. Institutional values are generally regarded as more important than instrumental values. He says:

> *The distinction is invidious if it is taken to mean that institutional values, are, by their nature, higher. Nothing of this sort is here implied, and the distinction is empirical: any value which organizes, directs, and integrates other values is, in respect to those other values, higher.*[17]

Looking at American values as such may extend a way to reconcile the black observations with the white. Whites feel that the American Creed and the ideals therein, the Protestant ethic and the Judeo-Christian ethos are highly valued and that the race problem in this country is caused by the moral conflict engendered by their violation. Blacks feel that white supremacy, European superiority, male superiority, and the superiority of people with money are highly valued, and that the race problem in this country is caused by the white rage incited by their rejection. It would seem, then, that there are two sets of values. Could one be institutional and the other instrumental? If so, which one organizes, directs, and integrates the other?

Instrumental values are derived from tool-using, observation, and experimentation. They are temporal, matter-of-fact, and secular. According to this definition, equality, liberty, fraternity, and justice would be instrumental values. White supremacy, European superiority, male superiority, and the superiority of people with money would be the institutional values. Every institution supports these values in practice and they are defended by force. Wheelis observes the following:

> *Institutional values derive from the activities associated with myth, mores, and status. The choice involved purports to be final. Such values do not refer to, but transcend, the evidence at hand. They claim absolute status and immunity to change, but are, in fact, relative to the culture that supports them: Christian sacraments are without validity in India, and suttee has achieved no validity in the western world. The final authority of such values is force.*[18]

On the other hand, instrumental values do not transcend the evidence at hand, but derive from progressively refined attention to such evidence. They possess transcultural validity.

> *They are, however, relative to the state of empirical knowledge at any given time, and change as that knowledge is enlarged. The final authority of such values is reason.*[19]

The confusion in interpretation is derived from observations.

It would appear that equality would be as good for a black

man as it is for a white man, and it is. Yet the concept of white superiority organizes this value so that the former transcends the reality of the latter. Consequently, there is a white view and a black view of the bridge. Berger and Luckmann call this a clash of symbolic universes. They say:

> *The intrinsic problem becomes accentuated if deviant versions of the symbolic universe come to be shared by groups of "inhabitants." In that case, for reasons evident in the nature of objectivation, the deviant version congeals into a reality in its own right, which, by its existence within the society, challenges the reality status of the symbolic universe as originally constituted. The group that has objectivated this deviant reality becomes the carrier of an alternative definition of reality.*[20]

The black alternative definition is more ascriptive than achievement-oriented. In fact, human beings could be arranged on a continuum and each be given a place.

Those people possessing the institutional values would possess or strive for the instrumental values. Those not possessing the institutional values would not possess the instrumental values, and, possibly, not strive for them. Arranged on a continuum, the highest order human being would be the white man of European descent with money (see Figure 1). The lowest order human being in the male universe is the black man with no money. The highest order human being in the female universe is the white woman with money, and the lowest order human being in the social order is the black woman with no money.

Such information raises many questions. Which value is prime? What about age and religion? Will sex guarantee more participation, equality, liberty, or fraternity than race? Do the two points of the black man with money really equal the two points of the white man without money? Do black men marry white women because the black woman is despised by all other humans in the universe, him included? Why do people say the white man and the black woman are the two free people in the universe, when he is the most preferred and she is the most despised? All such questions could lead to hypotheses for future research.

Male superiority	White European superiority	Money	Points
+	+	+	3
+	+	−	2
+	−	+	2
+	−	−	1
−	+	+	2
−	+	−	1
−	−	+	1
−	−	−	0

Figure 1. Ascribed positions according to value of humans.

THE SOCIAL SCIENCE OF ERROR

One function of social science has been to legitimatize the values of the white European culture by producing for distribution and dissemination a body of knowledge supportive of a model which guarantees success to white male Europeans.[21] Although social science's goal is truth, it has largely contributed to error. Yankelovich and Barrett describe the dilemma in this way:

> *If a scientific theory is a scheme of abstractions analogous in function to a map, then its truth (independently of its utility) lies in how well it depicts certain features of reality when reality's full concreteness and density can never be wholly captured by means of concepts, since abstractions, by definition, exclude full concreteness.*[22]

While western modes of knowing present clear and distinct abstractions, they obscure some aspects of reality.

Insight, understanding, empathy, intuition, prehension, and common sense are neglected.

> *The main reason for their neglect is the difficulty of verifying intuitive knowledge by those "objective" methods which are the cornerstones of modern science. Forms of knowledge which do not lend themselves to verification by such methods are minimized or overlooked entirely.*[23]

Glaser and Strauss press for theory which emerges from the data, instead of wasting time and good men in attempts to fit a theory based on reified ideas of culture and social studies. This is one of few attempts at the invention and codification of new methods for verifying insights.[24]

Such pursuits, hopefully, will lead to the search for truth. Knowledge, so produced, would be the power to set men free from oppression. The power of knowledge to do so may be investigated with an anthropological formula introduced by Sol Tax.* According to this formula, the groups supported by the value system (men, white Europeans, and the moneyed) are A groups (groups with power), and the groups deprived by the values (women, blacks and nonwhites, non-Europeans, and the poor) are B groups (groups without power). A has power over B.

Power derives from control of the land and its resources. It, then, is the ability to influence the behavior of others and to make them do what they may not want to do. A universe-maintenance system must preserve the status quo to keep these groups in power. Berger and Luckmann argue that this is done through the control of knowledge. They explain it in this manner:

*This formula was introduced by Sol Tax in Exhibit 44, "The Freedom To Make Mistakes," in *Documentary History of the Fox Project*, Fred Gearing, Robert McNetting, and Lisa R. Peattie, eds. (Chicago: University of Chicago, 1960), pp. 245-250.

> *Knowledge in this sense, is at the heart of the fundamental dialectic of society. It "programs" the channels in which externalization produces an objective world. It objectifies this world through language and the cognitive apparatus based on language, that is, it orders it into objects to be apprehended as reality. It is internalized again as objectively valid truth in the course of socialization.*[25]

If power groups decide what can be revealed and what can be hidden, A groups create reality for B groups and B groups hold inaccurate conceptual maps of reality.

Yankelovich and Barrett, once more comparing maps and theories, say that both may be (a) useless and false, (b) useless and true, (c) useful and false, and (d) useful and true.[26] Blacks feel that social science research about blacks has more often than not been useful and false for whites and useless and false for blacks. A short review of several studies will illustrate the use of conceptual truths or truths of abstraction as vehicles for the creation of inaccurate conceptual maps.

One well-known study which centers around the truth of abstractions and the method of verification is Daniel P. Moynihan's study of the Negro family.[27] The abstraction, pathology, pervaded the report. Moynihan, influenced by Catholic welfare philosophy, the failure of the War on Poverty community action programs, and his work with Nathan Glazer on the black family, assumed that the black society was in the process of deterioration at the heart of which was the black family. Furthermore, he assumed that that family was caught up in a tangle of pathology which perpetuated itself through a vicious cycle which could be broken when the distortions in the black family life were corrected.[28]

As evidence in his verification process Moynihan presents the following data: (a) nearly a quarter of urban black marriages are dissolved; (b) nearly a quarter of black births are illegitimate; (c) nearly a quarter of black families are headed by females; and (d) the breakdown of the black family has led to a startling increase in welfare dependency. He buttresses these data with several observations which develop his argument. He says that the roots of the problems are to be found in slavery, the position of the black man, urbanization, poverty, unemployment, and inadequate wages. Moynihan sees the problems increasing because of the increase in the black population.

He also suggests that the structure of the black family life is weakened by the pathology of the matriarchy, the failure cycle of youth, crime rates and youth delinquency, the low position of the black wage earner in the job market, and the effects of drug addiction. He concludes that "a national effort towards the problem of black Americans must be directed towards the question of family structure." The object should be to strengthen the black family so as to enable it to raise and support its members as do other families.

Moynihan could have as well reported the facts in a more positive light to demonstrate that black families have maintained a remarkable measure of stability despite the discrimination against blacks and females. It is significant, furthermore, that black people today can point to records which show that a majority of their marriages are stable, that three-fourths of their births are legitimate, and that males do head three-fourths of their families.[29] He could have mentioned that forced marriages and abortions in the black community are rare because of the economic instability, and this fact might affect the disparity between legitimate and illegitimate births among blacks and whites.

The astonishing result is that Moynihan clearly assigns the responsibility for the existing problems in the black community to the deteriorating black family, when the major factors that he, himself, uncovers occur outside of the family structure. It is not within the power of the black community to combat on a significant scale the mass discrimination practiced against it in the areas of salary, occupation, and general employment. In order to insure fair and equal job opportunities for black Americans, discrimination in employment must be attacked on a national level with vigor. Consequently, Moynihan's report to blacks is useless and false.

Another study conducted by Gunnar Myrdal used the abstraction, guilt. Myrdal, a Swedish social economist, imbued with the ideas of the Swedish welfare state which is noted for its socialistic liberality and where there is no noted race problem, used the Americans' Creed as an example of American ideals. He called the disparity between these ideals and American behavior the American dilemma.[30] A dilemma is a situation requiring a choice between equally undesirable alternatives or it is a difficult situation or perplexing problem. Myrdal does not make it clear which one of these definitions he uses. His central thesis is that a moral conflict exists in the hearts of

Americans because the treatment of the Negro violates the Americans' Creed. He assumes that Americans accept blacks as human beings therefore worthy of equality, liberty, fraternity, justice, and the pursuit of happiness. This is his major problem.

Myrdal studies the victim but not the bullet which killed him, nor does he attend adequately to the value system of the culture which produced the murderer. The value of white superiority and the greatness of western European civilization is buried in the doctrine of antiamalgamation wherein the primary valuation is racial purity.[31] But there is no entry in the index for European superiority.

Most seriously, Myrdal establishes no scale to measure his bias: friendliness to European civilization and white superiority. Nor does he provide a scale for his bias toward assimilation as a solution to the problems of black people who are oppressed. Additionally, he presents the principle of cumulation as causal, which it is not.

Myrdal's principle of cumulation says: (a) as discrimination eases, things for blacks will get better; and (b) as discrimination elevates, things will get worse. Using cumulation Myrdal is free to postulate that blacks can do little to help themselves unless whites approve. In fact, he discourages the black initiative as do present-day black integrationists[32] who infer that the moral conflict in the hearts of whites must effect change, and that blacks should wait for this alteration. In fact, Myrdal argues against protest and black nationalism.[33]

The principle of cumulation used in this study excludes synergism and epigenesis. It is not highly selective because Myrdal failed to reinforce his own criteria. His criteria are explicit and he says:

> *The value premises should be selected by the criterion of relevance and significance to the culture under study. Alternative sets of value premises would be most appropriate. If for reasons of practicability only one set of values is utilized, it is the more important that the reservation is always kept conscious: that the practical conclusion—and, to an extent, the direction of research—have only hypothetical validity and that the selection of another set of value premises might change both.*[34]

Obviously, Myrdal is operating on a hidden set of values which lead him to observe that accommodation is undoubtedly stronger than protest, "although drives for assimilation will generate protest, particularly in the South where the structure of caste is pervasive and unyielding," another myth. Today it is known that all America is the South. As Preston Wilcox puts it, there is "up South" and "down South."[35] Myrdal's study is in trouble where explanation is concerned, making it useless and false.

The study probably best known for its notoriety is the work of Arthur Jensen. His study is based on a nonoperating definition which he insists is testable. Quoted in *Life* he says:

> *Intelligence, like electricity, is easier to measure than to define.*[36]

Jensen simply defines intelligence as whatever the IQ tests measure. Armed with data therefrom he sets out to show that blacks, as a population, score significantly lower on IQ tests than the white population and attributes these lower IQs to genetic heritage, not to discrimination, poor diet, bad living conditions, or inferior schools. His study, published in the prestigious *Harvard Educational Review*, Winter 1969, raised a furor in the black community because of its dehumanizing and derogatory implications. Some scientists indicated that Jensen had not proved his proposition in a scientifically acceptable way. Others defended the assertion that science could not begin to differentiate hereditary variations in intelligence from environmental until social conditions had been equal for both races for several generations.[37] Jensen deals with the old nature-nurture argument but without redefining the problem or developing a broad metatheory. Jensen's work is useless and false.

Edward C. Banfield, high priest of political science, reigning as the Henry Lee Shattuck Professor of Urban Government at Harvard University, and head of President Nixon's task force on model cities, has produced another book. The author does not claim this study as a work of social science. He describes it this way:

> *Although I draw on work in economics, sociology, political science, psychology, history, planning, and other fields, this*

> *book is not really a work of social science. Rather, it is an attempt by a social scientist to think about the problems of the cities in the light of scholarly findings.*[38]

In view of the problems of social science discussed above, and the review of works contributing to error, one would expect to have problems with Banfield's explanations and solutions.

His abstraction is the normal culture which, of course, is White Anglo-Saxon Protestant with the prime acculturating variable, the postponement of gratification or future orientation. Banfield's central thesis is false when applied to blacks because the lower-class black culture is future-oriented. If one is good on earth and adheres to the rule of God, one will surely go to Glory! If this is not a manifestation of the postponement of gratification, what is?

Banfield has definite biases toward upper middle-class and middle-class people. He states that they plan ahead for their children. He attributes this to their "future-orientation" not to their possession of resources. They are "self-respecting" he says, "self-confident, and self-sufficient."[39] The final affront is the assertion that upper-class people abhor violence.[40] The fact that police are subsidized by the rich as are armies and wars in order that they may make money and keep it, is certainly contradictory. The solutions emanating from Banfield's unscholarly analysis are disastrous. Some of them are incorporated in the new unconstitutional crime legislation for Washington, D.C.[41]

The full meaning of these solutions are not readily understood because Americans receive little instruction on capitalism; consequently, it is often misinterpreted. Based on a competitive model of contrient interdependence, capitalism will always have losers. Deutsch explains this situation:

> *. . . "Contrient interdependence" is the condition in which individuals are so linked together that there is a negative correlation between their goal attainments. The degree of contrient interdependence between two individuals refers to the amount of negative correlation; it can vary in value from 0 to -1. In the limiting case, under complete contrient interdependence an individual can attain his goal if and only if the others with whom he is linked cannot attain their goals.*[42]

Said another way, when A wins, B loses, and when B wins, A loses. Then a group which is winning strives to keep that position by excluding another group.

Additionally, the economic paradigm of supply and demand operates in exactly this way. Parsons describes this adequately in his discussion of inclusion in the economy:

> *There are demands for inclusion–both from the excluded group and from certain elements who are already "in" and there is a supply, which also operates on both sides of the exclusion line. Supply here refers, for the excluded groups, to their qualifications for a membership, a matter of their cultural and social structures.*[43]

He makes it clear that the group already "in" determines the qualifications thereby controlling the supply. The qualifications demand acculturation and accommodation.

Parsons further describes the process in this manner to clarify the losers:

> *. . . On the side of the receiving community, "supply" consists in structural conditions which create institutionalized "slots" into which the newly received elements can fit, slots structured in accordance with the basic citizenship patterns of the developing community, not opportunities for crude "exploitation" by its members. Supply in this sense refers to a set of structural conditions on both sides of the "equation."*[44]

Here Parsons' biases obstruct a clear description. His comments about "exploitations" are not compatible with "slots structured in accordance with the basic citizenship patterns of the developing community" nor are these comments congruent with contrient interdependence as discussed by Deutsch. For the group already "in" determines not only the supply but the demand. It decides who will work, when, and where.

The economic institutions provide the mechanism for confining blacks to the kind of participation in the competitive model wherein A can continue to win and B can continue to lose. With money the prime determinant of life and happiness, it is highly unlikely that A groups will willingly, without bloodshed, relinquish their rights to certain jobs, career opportunities, resources, and capital.

Lastly, the philosophical premises of historians and anthropologists preserve the universe-maintenance system of the white European culture and its values. In history a temporal-spatial arrangement of facts occurs on a white European time line into which all other facts must be squeezed. For example, a fact is that Columbus discovered America. The question asked is how could this be a fact since people were already here and occupying the country. Obviously, someone else discovered America. Who were they? White European values preclude a vigorous search for this answer.

In anthropology, the hard work of L. S. B. Leakey has failed to receive the accolades that have been the rewards of other anthropologists. His diggings in Olduvai Gorge, Kenya, in East Africa have been largely self-financed and the academic community has begrudgingly acknowledged these findings. Should he prove that man began in Africa would white European values be diminished?[45]

Social science serves to legitimatize sets of knowledge for maintaining the status quo. The social scientists then design models for achieving conformity to the rules, regulations, laws, and standards which are necessary for that maintenance.

SOCIAL SCIENCE AND EDUCATION

Conformity is attained through institutionalization, legitimation, socialization, internalization, and identity. These processes occur through institutions. Most social scientists study institutions. A defining characteristic of an institution is a set of rules or norms. In fact, an institution can exist only when its rules are obeyed.[46] For black folk the school is the prime vehicle for institutionalization and socialization. Education is firmly entrenched in the social sciences, and the question of conformity is central. Jules Henry deplores this preoccupation here:

> *All educational systems aim at a steady state—a condition in which, on the one hand, the system tirelessly corrects deviations from the prescriptions of the culture, while on*

the other hand, the corrections become part of the psychoneurological equipment of the child and ultimately of the adult.[47]

Henry defines present education as socialization.

It is used in the broadest sense meaning "the whole process by which a newborn infant becomes a member of society, a member of his particular society, and an individual in his own right."[48] The concept of culture as a determinant of this process and of the norms and values to be perpetrated led to the "cultural deprivation" theories which were to account for the "deviants from the cultural prescriptions."

There is still controversy over the definition of culture. Valentine gives two definitions, the classical and the modern:

> *. . . that complex whole which includes knowledge, belief, art, morals, laws, customs, and any other capabilities and habits acquired by man as a member of society.*
>
> *. . . the organization of experience shared by members of a community, including their standards for perceiving, predicting, judging, and acting.*[49]

Valentine argues that social anthropologists treat the culture concept in ways that are different from the cultural method. They are generally more interested in social relations and social institutions rather than culture and use culture as a synonym for society.[50]

Education is the process which achieves conformity to the cultural values. Through it, explanations are developed about the social life of humans. Kluckhohn describes the process:

> *Social life among humans never occurs without a system of "conventional understandings" that are transmitted more or less intact from generation to generation. Any individual is familiar with some of these, and they constitute a set of standards against which he judges himself. To the extent that he fails to conform he experiences discomfort, because the intimate conditioning of infancy and childhood put great pressure on him to internalize these norms and his*

> *unconscious tendency is to associate withdrawal of love and protection or active punishment with deviation.*[51]

The school, then, becomes another place for the inculcation of the values or the large-ended goal statements of the society.

These are derived from the culture which develops from the relationship of the people to the land and the knowledge they accrue about that relationship. Norms are the conventional standards, the rules, regulations, and the laws which a person must judge himself against, comply with, and obey in order to uphold these values and maintain his group's relationship to the land. There is a system for internalizing the objective world for individual motivation to comply and obey and there is a system for objectifying reality for this internalization. Identity is the ultimate product of these processes. This is depicted in Figure 2.

Wheelis posits that identity is based on values especially those which are at the top of the hierarchy of beliefs, faiths, and ideals which integrate and determine subordinate values.[52] He defines identity further as a coherent sense of self, elaborating in this manner:

> *It depends upon the awareness that one's endeavors and one's life make sense, that they are meaningful in the context in which life is lived. It depends upon stable values, and upon the conviction that one's actions and values are harmoniously related. It is a sense of wholeness, of integration, of knowing what is right and what is wrong and of being able to choose.*[53]

Erikson contributes more to the definition of identity.

The term is probably more closely associated with him than with any other social scientist. He discusses identity in terms of three definitions: (a) group identity which is the group's basic ways of organizing experience which is transmitted to the infant's early bodily experiences and, through them, to the beginnings of his ego; (b) the personal identity which is the perception of the selfsameness and continuity of one's existence in time and space and the perception of the fact that others recognize one's selfsameness and continuity; and (c)

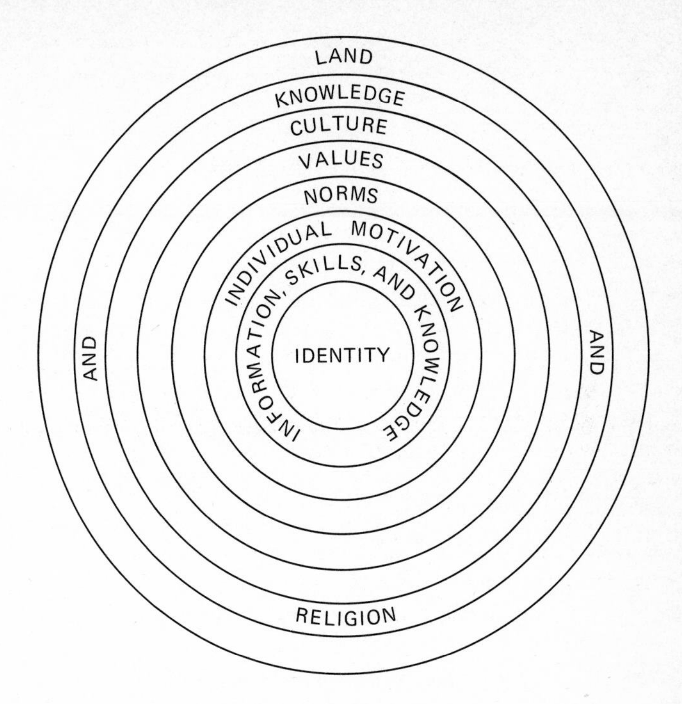

Figure 2. System of universe-maintenance.

ego identity which is the awareness of the fact that there is a selfsameness and continuity to the ego's synthesizing methods, the style of one's individuality, and that this style coincides with the sameness and continuity of one's meaning for significant others in the immediate community. The progress of growth from group identity to ego identity occurs through epigenesis discussed earlier.[54]

Expressed still another way, the self is the total potentiality of one's beingness. This potential is bombarded by the culture. The new structure which emerges from this bombardment results in the development of personal relationships and is often labeled the personality. When the person is able to identify other than self and what is personal for self within the cultural climate, the self-concept is born. The self-concept is an aggregate of roles which gives power for the attainment of identity. During the selection of roles, the person defines his

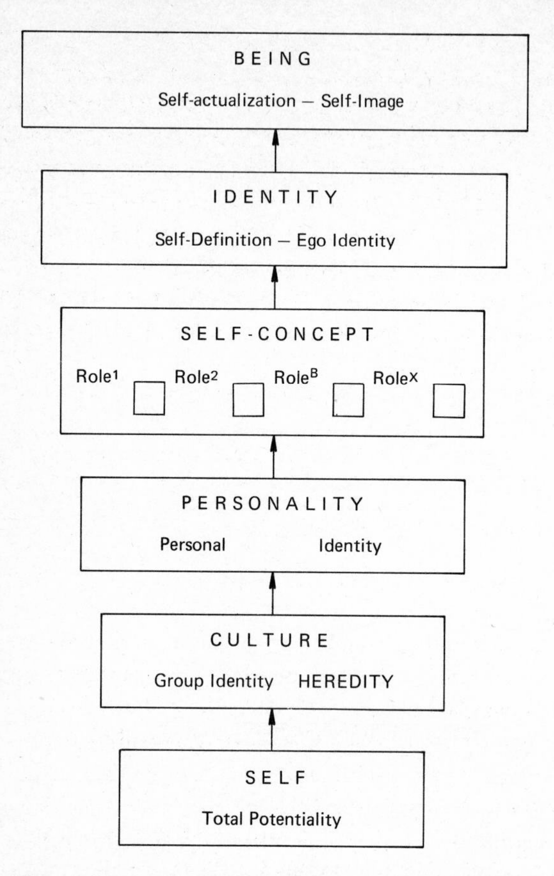

Figure 3. The process of being. This theoretical framework and conceptual map was developed by Edythe Stanford Williams, Director of Follow-Through Ethno-Linguistic Cultural Model, Center for Inner City Studies, Northeastern Illinois State College.

limits of freedom and uses the personality to express the self fully. The successful achievement of self-definition and ego-definition permits the actualization of the self (one's potential). The actualization of this potential is the self-image which encompasses everything of value to the individual (see Figure 3).

Society's role is to provide institutionalized settings for the experiences vital to the achievement of self-definition. Erikson argues that the adolescent needs support to help him to maintain powerful ego defenses against a growing intensity of impulse; to align his most important conflict-free achievements with work opportunities; and to resynthesize childhood development in a way that accords with roles offered by the society.[55] Yankelovich and Barrett agree and argue persuasively that the self is a whole and must not be defined by reductionism, that consciousness has to be restored to a more central role, and that the individual does have some degree of control over his life although harshly limited.[56]

EDUCATION AND A BLACK IDENTITY

Today black people are defining themselves. Having discarded the meaningless name, Negro, they are crying, "I'm black and I'm proud." The demands for Black Literature and Black History are attempts at restructuring the temporal-spatial arrangements of history to accommodate the presence and past of other pseudospecies, to remove the constraints on the dissemination and distribution of knowledge, and to permit the installation of the worth of blackness as a value in the black community.

Natural Afro hair styles, dashikis, "uhuru," and "umoja" are symbols of this value-building effort. The idiom has been dignified by the new poets who are nation-building. Brother and sister are common terms of address after the Community of Islam. A new goal, liberation, has been substituted for integration or racial balance. The new black is saying that he will be included, liberated, with full preservation of his ethnic and racial differences, and he will not hate himself or turn white to do so.

Blacks, now, are electing to reject the norms, one of the three routes prescribed by Goffman for stigmatized groups.[57] The three routes are:

1. To accept the norm of black inferiority which upholds the value of white superiority, but to refuse to be defined by it as an individual. Such a person would say that blacks

are inferior, dirty, lazy, "but not me!" "I obey the rules, regulations, and laws. I am properly motivated to conform and I have internalized all the information, skills, and knowledge. I made mine." This person is acculturated and accommodates to the alien cultural values. Acculturation is the process of adopting the cultural traits or social patterns of a group other than one's own.

2. To reject the norm of black inferiority. Such a person would say, "Black is beautiful. Black people have a proud and distinguished African heritage. I will not meet standards, norms, obey laws, rules and regulations which dehumanize my people. I will not internalize information, skills, and knowledge which deprive me of my humanity, identity, and culture." Such a person is enculturated. He protests the culture from which alien values emanate. Enculturation is the process by which a person adapts to his own culture and assimilates its values.[58]

3. To pass and to cover. Such a person gives up or surrenders his right to be black and becomes white. This is assimilation. Education for the black identity must deal with enculturation processes.

Allison Davis and the Gardners looked at these processes in their study, *Deep South.*[59] They revealed that the worthlessness of blackness and the fear of white reprisal were the cornerstones of the caste system. Black children imbued with these ideas are docile and submissive. They grow up to hate themselves and others like them. The models they imitate are white. Those who refuse to do so are punished by the larger society. Consequently, most blacks opt for acculturation and accommodation. It has always been harder to be black and proud and unafraid of white retaliation. Education for a black identity must face the caste cornerstones and find an answer to this problem: how does one struggle hard enough to attain liberation but cautious enough to avoid extermination?

The environment is so oppressive that it is impossible for the black identity to flourish. This oppression prevents the provision of experiences necessary for the inborn structures of young blacks to develop according to their maturational schedules. Education for a black identity must make these provisions. Carter Woodson argues that the program for the uplift of the Negro must be based upon a scientific study from within "to develop him the power to do for himself what his oppressors will never do to elevate him to the level of others."[60] He urges the Negro teacher to treat the disease rather than the symptoms.

The disease is the value system. It must be replaced. Civilizations and cultures can persist only if the values are upheld. A truly revolutionary movement undermines the values. If the drive toward black pride continues and gains strength, the erosion of the cornerstone, worthlessness of blackness, may lend great impetus to liberation.

Power is needed to disobey the white cultural imperatives without severe punishment. Since blacks have no land, the only power is people. Previously excluded groups consolidated their people from a separatist vantage point with an ideology composed of: (a) a pseudo-species declaration which said "we are the chosen people of God"; (b) a territorial imperative which was a native land or point of origin; and (c) an identity specification, i.e., Irish-Catholic, Jewish, Muslim.

Separatism intensified the "we groupness" until a strong nationalism was developed. Then myths, rites, rituals were created to sustain it. Associations and organizations were developed to perpetuate it. This nationalism projected a negative identity which rejected all others as aliens and forced the people into a kind of "promotively interdependent" situation which Deutsch calls cooperation.[61] Trust then built an economic niche and a work bloc which the group dominated and thereby controlled an access to the economy. Once the capitalistic base was established, the group then vaulted into the political arena, effected coalitions, and worked for power. This process has been called a power-inclusion model.[62]

Except for the Community of Islam, blacks do not have the ideology necessary for a separatist vantage point. Education for a black identity must be concerned with the development of such an ideology. Berger and Luckmann describe ideology as a particular definition of reality which comes to be attached to a concrete power interest. They explain as follows:

> *Every group engaged in social conflict requires solidarity. Ideologies generate solidarity. The choice of a particular ideology is not necessarily based on its instrinsic theoretical elements, but may stem from a chance encounter.*[63]

	NATIONALISM				INTEGRATION			
	Violent	Non-violent	Accommo-dating	Pro-test	Violent	Non-violent	Accommo-dating	Pro-test
Panthers					X			X
S.C.L.C.						X		X
Islam		X	X					
RNA	X			X				
NAACP						X	X	

Figure 4. Ideologies and derivative strategies. (S.C.L.C. is Southern Christian Leadership Conference, RNA is Republic of New Africa. NAACP is the National Association For The Advancement of Colored People).

Although unity has been the goal of black people, an ideology has failed to emerge to achieve it.

The processes of enculturation and acculturation led to the development of two ideologies: integration and nationalism. Integration movements, generally speaking, are acculturated movements, and usually solicit white support. Nationalist movements, generally speaking, are enculturated movements, and are black supported. The ideology of the struggle for liberation has vacillated between nationalism and integration. Nationalism is the strong feeling of attachment to a nation. In the case of black folk, this nation is Africa.[64] Integration refers to the openness of society, to a condition in which every individual can make the maximum number of voluntary contacts with others without regard to qualifications of ancestry.[65] The strategies of these ideologies depicted in Figure 4 are violence, nonviolence, accommodation, and protest. All organizations use both accommodation and protest in some form. The chart points out the predominant alternative used by the organization.

Education for a black identity must provide a broader set of knowledge to provide the information and skills to construct alternative institutions for enculturation processes. The present emphasis on the African heritage and black culture will provide some experiences for the growth of this new black identity. But, the identity is never complete. It is open-ended and thereby subject to threat. The new black must discover ways to preserve the black identity or he may be forced to surrender it again. Presently, the white is dominant and defines reality. But his limits, too, are set by nature and nature acts back.

New structures emerge as humans interact with nature and the socially constructed world. In this process change occurs. In order to inject new cultural ideas and new direction for the pursuit of truth into present attempts at expanding the frontiers of human knowledge, blacks must invade the sacred precincts of the physical sciences from which they are being systematically excluded as well as the social sciences. Scientific inquiry can provide approaches useful to black people, and education must be concerned with that search.

Teachers and educators must unearth and disseminate knowledge, skills, and information which construct more accurate conceptual maps of reality for: (1) understanding the people blacks must unify; (2) identifying those who might oppose this unification and neutralizing or destroying their effects; (3) differentiating the real world from its illusion; and (4) providing the insights needed to discover alternatives otherwise unavailable. From these conceptual maps new synergistic structures will evolve: a strong cultural base for black people and an enculturated black identity based on new values.

NOTES

1. Henry A. Banks, "Black Consciousness: A Student Survey," *The Black Scholar* (September 1970), p. 44.
2. Imamu Ameer Baraka, "A Black Value System," *The Black Scholar* (November 1969), pp. 54-60.
3. This statement refers to the resistance against the research of Banfield, Jensen, and Moynihan discussed later in this chapter.
4. George C. Homans, *The Nature of Social Science* (New York: Harcourt, Brace & World, Harbinger Books, 1967), pp. 3-23.
5. Homans, op. cit., pp. 10-18.
6. Ibid., p. 18.

7. Ibid., pp. 106-107.
8. Daniel Yankelovich and William Barrett, *Ego and Instinct* (New York: Random House, 1970), pp. 177-448.
9. Yankelovich and Barrett, op. cit., p. 307.
10. Ibid., p. 329.
11. Ibid., pp. 434-435. See also: Erik H. Erikson, *Identity, Youth and Crisis* (New York: Norton, 1968), p. 92.
12. Erikson, op. cit., p. 93.
13. Yankelovich and Barrett, op. cit., p. 334.
14. Yankelovich and Barrett, op. cit., pp. 394-413.
15. Robert Rosenthal and Lenore Jacobsen, *Pygmalion in the Classroom* (New York: Holt, 1968), p. 38.
16. Yankelovich and Barrett, op. cit., p. 461.
17. Allen Wheelis, *The Quest for Identity* (New York: Norton, 1958), p. 182.
18. Wheelis, op. cit., p. 179.
19. Wheelis, op. cit., p. 179.
20. Peter L. Berger and Thomas Luckmann, *The Social Construction of Reality* (Garden City, N.Y.: Doubleday, 1966), p. 98.
21. Barbara A. Sizemore and Kymara S. Chase, "I Dig Your Thing, But It Ain't In My Bag," *Notre Dame Journal of Education* 1, no. 3 (Fall 1970).
22. Yankelovich and Barrett, op. cit., p. 339.
23. Ibid., p. 341.
24. Barney G. Glaser and Anselm L. Strauss, *The Discovery of Grounded Theory* (Chicago: Aldine, 1968), pp. 45-77.
25. Berger and Luckmann, op. cit., p. 62.
26. Yankelovich and Barrett, op. cit., p. 338.
27. Daniel P. Moynihan, *The Negro Family: A Case for National Action* (U.S. Department of Labor, 1965).
28. This review of Moynihan's report was taken from a paper by Barbara J. Neverdon, submitted in partial fulfillment of a course at Northeastern Illinois State College Center for Inner-City Studies.
29. For further emphasis of this point, see: Andrew Billingsley, *Black Families in White America* (Englewood Cliffs, N.J.: Prentice-Hall, Spectrum Book, 1968).
30. Gunnar Myrdal, *An American Dilemma* (New York: Harper & Row, 1962).
31. Ibid., pp. 53-57.

32. See: "Which Way Black America?" *Ebony* (Special Issue, August 1970) for comments by leading blacks.
33. Myrdal, op. cit., p. 749.
34. Ibid., p. 1045.
35. Preston Wilcox, "The Kids Will Decide," *Ebony* (August 1970), pp. 134-137.
36. John Neary, "A Scientist's Variations on a Disturbing Racial Theme," *Life* (June 12, 1970), p. 580.
37. Walter F. Bodmer and Luigi Luca Cavalli-Sforza, "Intelligence and Race," *Scientific American* 223, no. 4 (October 1970), pp. 19-29.
38. Edward C. Banfield, *The Unheavenly City* (Boston: Little, Brown, 1970), p. v.
39. Banfield, op. cit., p. 49.
40. Ibid.
41. For information on Washington, D.C., crime legislation, see *Chicago Sun-Times,* July 30, 1970, for full discussion.
42. Morton Deutsch, "Cooperation and Trust: Some Theoretical Notes," in *Interpersonal Dynamics,* Warren G. Bennis, Edgar H. Schein, David E. Berlew, and Fred I. Steele, eds. (Homewood, Ill.: Dorsey, 1964), p. 566.
43. Talcott Parsons, "Full Citizenship Rights for the Negro American," in *The Negro American,* Talcott Parsons and Kenneth B. Clark, eds. (Boston: Houghton Mifflin, 1965), pp. 721-722.
44. Parsons, op. cit., pp. 721-722.
45. L. S. B. Leakey, *White African* (Cambridge, Mass.: Schenkman, 1966), pp. 207-316.
46. Homans, op. cit., p. 50.
47. Jules Henry, "Is Education Possible? Are We Qualified to Enlighten Dissenters?" in *Public Controls for Non-Public Schools,* Donald A. Erickson, ed. (Chicago: University of Chicago, 1969), pp. 100-102.
48. John A. Clausen, "A Historical and Comparative View of Socialization Theory and Research," in *Socialization and Society,* John A. Clausen, ed. (Boston: Little, Brown, 1968), p. 41.
49. Charles A. Valentine, *Culture and Poverty* (Chicago: University of Chicago, 1968), p. 3.
50. Ibid., p. 4.
51. Clyde Kluckhohn, "The Concept of Culture," in *Culture and Behavior,* Richard Kluckhohn, ed. (New York: Free Press, 1962), p. 67.

52. Wheelis, op. cit., p. 200.
53. Ibid., p. 19.
54. Erickson, op. cit.
55. Yankelovich and Barrett, op. cit., p. 131.
56. Ibid., pp. 323-324.
57. Erving Goffman, *Stigma* (Englewood Cliffs, N.J.: Prentice-Hall, 1963).
58. Clausen, op. cit., p. 41.
59. Allison Davis, Burleigh and Mary Gardner, *Deep South* (Chicago: University of Chicago, Phoenix Books, 1941, 1965).
60. Carter G. Woodson, *The Miseducation of the Negro* (Washington, D.C.: Associated, 1933), pp. 144-156.
61. Deutsch, op. cit., pp. 565-566.
62. For a complete discussion of this model, see: Barbara A. Sizemore, "Separatism: A Reality Approach to Inclusion?" in *Racial Crisis in American Education,* Robert L. Green, ed. (Chicago: Follett Educational Corp., 1969), pp. 249-276.
63. Berger and Luckmann, op. cit., p. 114.
64. For an in-depth discussion of nationalism, see: Parsons, op. cit., pp. 710-716; and Harold Cruse, *The Crisis of the Negro Intellectual* (New York: William Morrow, 1967), pp. 329-332.
65. Oscar Handlin, "The Goals of Integration," in *The Negro-American* Talcott Parsons and Kenneth B. Clark, eds. (Boston: Houghton Mifflin, 1965), p. 661.

BIBLIOGRAPHY

1. Banfield, Edward C., *The Unheavenly City* (Boston: Little, Brown, 1970), p. v.
2. Banks, Henry A., "A Black Consciousness: A Student Survey," *The Black Scholar* (September 1970), p. 44.
3. Baraka, Imamu Ameer, "A Black Value System," *The Black Scholar* (November 1969), pp. 54-60.
4. Berger, Peter L. and Thomas Luckmann, *The Social Construction* (Garden City, New York: Doubleday, 1966), p. 98.
5. Billingsley, Andrew, *Black Families in White America* (Englewood Cliffs, N.J.: Prentice-Hall, Spectrum Book, 1968).

6. Bodmer, Walter F. and Luigi Luca Cavalli-Sforza, "Intelligence and Race," *Scientific American* 223, no. 4 (October 1970), pp. 19-29.
7. Clausen, John A., "A Historical and Comparative View of Socialization Theory and Research," in *Socialization and Society,* John A. Clausen, ed. (Boston: Little, Brown, 1968).
8. Cruse, Harold, *The Crisis of the Negro Intellectual* (New York: Morrow, 1967), pp. 329-332.
9. Davis, Allison, Burleigh and Mary Gardner, *Deep South* (Chicago: University of Chicago, Phoenix Books, 1941, 1965).
10. Deutsch, Morton, "Cooperation and Trust: Some Theoretical Notes," in *Interpersonal Dynamics,* Warren G. Bennis, Edgar H. Scheine, David E. Berlew, and Fred I. Steele, eds. (Homewood, Ill.: Dorsey, 1964), p. 566.
11. Erikson, Erik H., *Identity, Youth and Crisis* (New York: Norton, 1968), p. 92.
12. Glaser, Barney G. and Anselm L. Strauss, *The Discovery of Grounded Theory* (Chicago: Aldine, 1968), pp. 45-77.
13. Goffman, Erving, *Stigma* (Englewood Cliffs, N.J.: Prentice-Hall, 1963).
14. Henry, Jules, "Is Education Possible? Are We Qualified to Enlighten Dissenters?" in *Public Controls for Non-Public Schools,* Donald A. Erickson, ed. (Chicago: University of Chicago, 1969), pp. 100-102.
15. Homans, George C., *The Nature of Social Science* (New York: Harcourt, Brace & World, Harbinger Books, 1967), pp. 3-23.
16. Parsons, Talcott, "Full Citizenship Rights for the Negro American," in *The Negro American,* Talcott Parsons and Kenneth B. Clark, eds. (Boston: Houghton Mifflin, 1965), pp. 721-722.
17. Rosenthal, Robert and Lenore Jacobsen, *Pygmalion in the Classroom* (New York: Holt, 1968), p. 38.
18. Sizemore, Barbara A., "Separatism: A Reality Approach to Inclusion?" in *Racial Crisis in American Education,* Robert L. Green, ed. (Chicago: Follett Educational Corp., 1969), pp. 249-276.
19. Sizemore, Barbara A. and Kymara S. Chase, "I Dig Your Thing, But It Ain't In My Bag," *Notre Dame Journal of Education 1,* no. 3, Fall 1970.
20. Valentine, Charles A. *Culture and Poverty* (Chicago: University of Chicago, 1968), p. 3.

21. Wheelis, Allen, *The Quest for Identity* (New York: Norton, 1958), p. 182.
22. Wilcox, Preston, "The Kids Will Decide," *Ebony* (August 1970), pp. 134-137.
23. Woodson, Carter G., *The Miseducation of the Negro* (Washington, D.C.: Associated Publishers, 1933), pp. 144-156.
24. Yankelovich, Daniel and William Barrett, *Ego and Instinct* (New York: Random House, 1970), pp. 177-448.

8

Negro Self-Concept Reappraised

Jean Dresden Grambs

COLORED, NEGRO, OR BLACK?

In the 1950s and early 1960s when one traveled in the Northeast, Mid-Atlantic, or South to discuss problems of desegregation and intergration, it was important to find out first what was the preferred minority group term. Clearly in the South one used "Negro," with enunciated clarity on the final "o." The term "colored" was summarily rejected and with reason. Wherever one traveled there were the ubiquitous signs: "White"–"Colored."

In the Mid-Atlantic areas and in the Northeast, opinion was not as clear. To be a "person of color" was often preferred, and "Negro" was sometimes considered a term of dubious value. While city newspapers such as those in Baltimore and Washington used the term "Afro-American," there was little apparent interest in general public usage of that term.

Now there is a difference: the Black American has emerged. The demands made by formerly passive and silent Negro youth are now for "black studies." Few newspapers and journals today use any but the term "black" to refer to Negro concerns. The new slogans are Black Is Beautiful, and Black Power.

1964 was not very long ago, yet it was in that year that the United States Supreme Court had to make it clear that a black woman was not in contempt of court when she refused to answer questions when called Mary rather than the courtesy title of Miss or Mrs. and her last name.

Advances made by blacks in occupationally visible mobility have been documented in detail.[1] The pressure on all kinds of institutions to "produce" instant blacks in places of prestige and status has resulted in intensive recruiting of staff from black colleges to "white" universities, and the rapid advancement of black personnel in many areas. The results, while providing most sensitive institutions with token black personnel, has engendered anger on the part of black colleges[2] who have lost personnel, and hostility on the part of white colleagues who see persons promoted ahead of them who may or may not merit it—because they are black.[3] These frantic scramblings to provide a semblance of integration in the work force have not been accidental. The selective boycotting of bakeries, home fuel oil delivery firms, and department stores by black purchasers has caused alert businessmen to move swiftly to avoid a negative public image and/or bankruptcy.[4] The pressure on college campuses on the part of both white and black students has prompted a quick increase in the number of black students in predominantly white colleges, though the actual number is still small.[5]

The income gap has not, however, been appreciably reduced. Blacks and whites with comparable education do not have the same earning power, and most blacks are still concentrated in the lowest income levels.[6] Housing patterns still reflect white racism, forcing blacks to live in substandard housing at exorbitant rents and in segregated areas.[7] Craft unions are still resisting any massive opening of apprenticeships or membership to blacks, despite Federal pressure.[8] Even with comparable education, blacks receive less income at lower paying jobs than whites, and promotion to managerial or supervisory posts is still the exception rather than the rule.[9]

Again, the publicly announced massive efforts of Federal programs to aid the urban poor, primarily black population, has been less than overwhelming in impact. In fact, recent studies of Title I of

ESEA show that it was not used as planned, that is, to raise the educational level of the children of the poor.[10] The poor are not only as poor as before, but, as inflation increases, have become poorer.[11] The most significant item, however, is that now the black poor know just how poor they are.

The movement of concerned white liberals to alleviate black poverty and powerlessness has produced two distinct reactions: the proliferation of self help and community action groups[12] (government sponsored or otherwise), and aggressive black militancy, both verbal and physical.

Our purpose here is neither to evaluate nor analyze these particular social changes. Ample documentation and discussion exists and will continue to fill books and journals.[13] The question of interest to the educator is: *To what extent has the black revolution had an impact on the black child? If there has been a change, what is it, and what is the appropriate educational response?*

Black pressure for a bigger slice of the pie, which in some well-publicized instances was couched in terms of nonnegotiable demands, has produced the inevitalbe counterreaction by whites antagonistic to any move of blacks out of their powerless and separated lives. Extremist positions make good newspaper copy and television news. The public is getting more exposure than ever before to polarized views, both white and black. Of course, regional differences operate: there is greater coverage of white reactionary views in the South, while justified black demands are muted. In the North, West, and Middle West, black demands are considered "news." Outside the South one can also hear about moderate white support for the validity of that which militant black leaders state at the upper level of decibel tolerance.

With each shift in tactics and related public events, individuals and groups emerge to "tell it like it is." There are, of course, both the white version and the black version—and many intervals between the extremes of either groups. But whose view is *right?*

WHO SPEAKS FOR THE BLACKS?

There is doubt in the minds of many as to whether any white person (and I am white) can present an authentic view of the black experience.

The point is a good one and worth examining in the context of this essay. Scholars experience no great discomfort in delving into the history and culture of nations and places alien to them. Anthropology has built its scholarship around the analysis of unfamiliar cultures. But it is agreed that few Frenchmen would accept an American interpretation of French diplomatic history. But then, even Frenchmen don't agree among themselves on that subject.

Historians have always been accused of chauvinistic views of history—any history. Despite years of effort, there are great gaps among historians regarding the "true version" of events when national or regional interpretations of "the facts" are written.[14] It is instructive to see the ways other nations report events in the history of the United States.[15] Women and American Indians are disturbed by the way historians have written them in—or out—of American history.[16] The 1960s were marked by herculean efforts on the part of publishers and writers to put the blacks back into the American history taught to children and youth.[17]

Some blacks today say that no white person can ever be astute enough to assess accurately the black experience. In fact, the case is made that a good deal of the problem of the black in America is that white America has written his history, described his psychology, analyzed his social conditions, and told him what he was like and who he was.[18] Thus, the claim is made that only black historians can write black history, and only black psychologists can describe the black psyche. But there are degrees of blackness. *Black Rage*, for instance, written by two qualified black psychiatrists, is not accepted by some blacks.[19] In reviewing the book I found it to be extremely valuable and insightful. A black doctor reviewed it otherwise.[20]

The preceding discussion of who is competent to speak about the black condition today, is a reflection of black revolution itself. Reaction against the obvious distortions of the black experience over many centuries by white (and black) commentators has been to reject all whites in whatever capacity. As noted, however, this may not only be unwise but impossible.[21] I would argue that the view of the white, as well as the black, is clarified by having more than one observer since the white-black experience takes place in a milieu that includes both. It is imperative that both parties to the social exchange also exchange their views. In this context, then, each party can identify those experiences which provide insight into minority group responses, and relate them to the experiences of the black child—or to any other person who is not like oneself.

HOW FARE BLACK YOUTH?

It is not an easy task to assess the impact of recent events on black youth. Research efforts are typically several years in process, so that one is usually acquiring information about a group or a program which may already be irrelevant and outdated. One suspects that intervening events may indeed cast doubt upon much research data on black and/or Negro children and youth.

Past studies of black youth, however, provide some historical perspective. Compared with contemporary literature there was more depth of study and research on black youth in the mid-forties than there is today. Such studies are often ignored or dismissed; for example, the many volumes in the intergroup education project headed by Hilda Taba,[22] as well as the studies of Negro youth sponsored by the American Council on Education which were eloquent, passionate, and revealing.[23] The only attempt at an experimental study of how interracial materials read to children could effect their out-of-class behavior, Trager and Yarrow's *They Learn What They Live,*[24] was published in 1952, and is seldom acknowledged in current research or reviews of the literature.

The research volumes noted above were part of the educator's fare in the 1940s. Few students in graduate education at that time were not, in some way or another, exposed to one or more volumes. The related classic problems in education, such as discrimination on the basis of social class (1942)[25] the inequities of ability grouping (1931),[26] and the granddaddy of school dropout studies, *Laggards in Our Schools* (1909),[27] pointed out the ways in which schools were dispensing unequal educational opportunity. In other areas—housing and employment—studies showed the patterns of American discrimination.[28] *The Strange Career of Jim Crow* first appeared in 1955, yet has only recently reached a wide audience.[29]

The intervention of war crises (World War II, the Korean war) appeared to deflect the impact of these findings. The demise of the Progressive Education Association, the rise of McCarthyism and attacks on the schools for supposedly subversive teaching,[30] also served to turn aside educational attention from the social problems facing schools. One should not, therefore, be surprised by the seeming naiveté of colleagues who suddenly 'discover' the facts of discrimination and the impact of segregated living, the subtle but innumerable slights and rebuffs experienced daily by black Americans. It must also be remembered that educators, too, were denied an opportunity to learn the complete history of our country. When talking with a group of black and white teachers in Birmingham, Alabama, prior to the first efforts to integrate the schools in 1968, I read passages from a history of the South which reported in detail the prevailing fear of slave uprising prior to the Civil War.[31] After I finished my presentation I was surrounded by teachers of both groups who wanted to know my source. They had no knowledge of these events in their own state history. On a number of occasions, when discussing problems of desegregation before school audiences, I have read the deleted passage regarding slavery, which Jefferson had drafted in the original list of complaints against the King of England for inclusion in the Declaration of Independence.[32] In only one or two instances did any teacher, among thousands, correctly identify the passage. The problem educators face, then, in assessing the impact of current events on today's black youth is compounded by this gap in informed and useful data. A generation of educators have to be re-educated.

The Supreme Court decision in 1954 outlawing de facto school segregation changed many things. It was fascinating to observe the rash of publications about the newly discovered "disadvantaged child." Despite hundreds of years of educating such individuals in the segregated systems of the South or the ghettos of the North, no one apparently had realized that these children were disadvantaged. What became critically apparent was that such children were not "culturally deprived" (one particularly offensive euphemism)—they had ample "culture." They were, in fact, educational cripples. Substandard schooling for generations, and substandard nutrition, and thwarted ambition had produced a phenomenon that white educators suddenly "discovered."

The urgency to know more about the problem of educating the child of poverty, typically perceived as a problem of the black inner-city pupil, has produced a plethora of "instant books." The list is extensive, and there is no room to review the material here.[33] The merit of the material thus produced is variable. The rush to find out what to do has, inevitably, also produced hurried prescriptions and self-laudatory success stories. Only by examining the assumptions made about how the institution effects the learner, as we are trying to do here, will we find that these new materials, guides, exhortations, and programs have any possibility for lasting impact.

Why were children who were segregated in schools or in life so much more difficult, seemingly, to reach and teach?

I had the temerity a number of years ago to try to explain (mostly for myself) what was peculiarly *different* about the educational problem of the Negro in America.[34] I saw it, in the Fall of 1963, as a problem in the development of the concept of the self. I hypothesized that, since a person's self was the screen through which he monitored messages coming in, and relayed messages going out, that one who felt good about himself conveyed this to others; and read most messages coming to him as reconfirming his own sense of his self-esteem. Conversely, persons who thought poorly of themselves indicated to others that this was their own self-valuation, and—as in the self-fulfilling prophecy—read messages conveyed to them as, again, affirming their lack of worth. I posited, in 1963, that part of the problem of lack of Negro achievement in school was due to a continual bombardment from society of messages conveying demeaning versions of who Negroes were. These messages were overtly spelled out in segregated schooling, housing, job discrimination, and the drastic repression of all aggression against such social displacment. The Negro could hate, but he could never let anyone else (white, that is) know it or feel it.

The hypothesis further went on to state that after being told for generations that one is not good, one begins to believe the myth. Thus, unto the fourth and fifth generation a sense of inadequacy, of not being really capable of achieving, of deserving the low place on the scale one was born into, becomes a barrier to benign intervention. Children of parents defeated by the schools are apt to expect similar treatment, because their parents have told them defeat is to be expected—and so they are.

Such a cynical cycle of defeat, I felt (in 1963) was educationally indefensible. We should have ways of stopping this vicious repetition of failure. And I proposed a few possible educational interventions that seemed to be more promising than others.

As the 1970s loom ahead, I find it a different, and difficult task to consider the sources and affects of school and culture on black children's views of themselves. The very words must change: I cannot talk any more about Negro self-concept: it is black self-concept. The shift in adjective is significant.

The discovery of the culturally disadvantaged child in the years between 1960 and 1970 has not all been a blessing. The recognition that some children came to school with a conviction that they could not learn has been reinforced by volumes and articles which identified all the ways in which such children are presumed to be hampered. The results have been not only to support the idea of one large group of children as being "different," but also to relieve educators of the responsibility for achieving educational success with such a poorly endowed product. A perceptive report of a study of the black students (in one major Midwestern university) who are provided special tutoring and financial assistance to help them survive in the predominantly white environment, points out the demeaning features of the Noble Savage attitudes which the black ascribe to the program and its sponsors. Dissatisfaction with the program, despite its success in keeping black students in the university, is due to the paternalism and hypocrisy which the black students perceive, notably because "The university community has subtly defamed blacks by failing to consult them regarding the structure and operation of what is really *their* program."[35]

There are some realities, however, which are tangible environmental factors influencing the achievement orientation of children and youth. One very obvious factor is that poverty itself tends to produce generational cycles of despair and defeat.[36] As Sarason noted, one child who was in trouble with juvenile authorities and doing poorly in school as well, was a member of a family known continuously by welfare service agencies since 1904.[37] All of the available help had been insufficient to get this family into a position where its members could cope with life's demands.

The educational potential a child is born with may be one already disadvantaged because of poor nutrition. A careful study of children who suffered severe malnutrition requiring hospitalization were studied along with their siblings who had not been hospitalized.[38] Those who had suffered severe malnutrition as a group were markedly inferior in school performance, and even though they were rehabilitated, scored in the bottom ranks of mental ability. As one of the researchers commented, "These 'insults of environment' . . . are lasting. They affect the child throughout life. And the next generation . . . ."[39] The undernourished child becomes a mother whose childhood malnutrition affects her own children. It is estimated that it takes at least three well nourished generations to eradicate the affects of severe malnutrition. Many of the children, therefore, who are perceived to be educationally inadequate are factually so. To what extent educational intervention can make a difference is questionable: the best that one could say is that schools might pay considerably more attention to the nutrition they teach, and which they provide via school breakfast, brunch, and lunch programs.

Even these, however, may be insufficient. As Greene and Ryan observed, the lack of awareness by school personnel of the eating habits and needs of the children leads to tremendous food wastage in one of the most impoverished areas of New York City.[40] Dietitions who make up school lunch programs seemingly make no effort to find out the eating habits of poverty children. The World Health Organization has grappled with the problem of inducing populations to shift diets so that their mental acuity as well as their health could be improved, but found that the new diet had to resemble in appearance, flavor and texture that diet to which the cultural group was attuned. Changes in food habits are almost as difficult to change as religious beliefs—and some are tied together. Yet, with our knowledge of nutrition and its impact on intelligence and school performance, educators blithely feed lunches to children who will not eat them. Equally nutritious food is available, if prepared and offered in the style that the group finds acceptable. One must hasten to add that, on the basis of personal visitations to the lunchrooms and cafeterias of many schools, it is surprising that anyone eats anything that is served. No matter how good the food; the noise, rush, general chaos, the dingy and sometimes dirty surroundings (have you ever been in a large school cafeteria at the end of the fourth or fifth lunch period?) are enough to discourage even the hungriest child or adolescent.

Public dismay over the inaction of governmental agencies to

correct the incidence of lead poisoning among slum children is another symptom of how a controllable factor can cloud perception of a whole array. Lead poisoning, which occurs in children who nibble peeling paint from old walls, is endemic in old, poorly maintained dwellings. It is typically a disease of poverty. And if it is not diagnosed early and treated quickly, it results in mental retardation or death. Health authorities in big cities know this, yet have been seriously derelict in providing the screening which could detect and then ameliorate the condition. Housing inspectors have not enforced codes which would eliminate lead based painted walls.[41] Lead-based paint on children's toys is still used. And it is the poor who suffer.[42]

The reality, then, of groups of children who are victims of the poverty they were born into, must be considered carefully by the educator. That some children in an inner-city classroom are suffering from the deprivation of malnutrition is clear; but the majority of the children *are* adequately fed and cared for. As Leacock points out, despite the low-income level of the children studied, most of them came from intact homes. Although they lived in apartment houses or housing projects, and although their families were large, including other relatives or boarders, "Four out of every five children interviewed in the study were living with both their mothers and fathers."[43] Again, the reality of the very desperate becomes a mythology which traps all. There are many black families with the mother as head of the household, but there are many more where there is a stable and "standard" family.

The classrooms of the inner city which are increasingly black, include children who are maimed by social pathologies—malnutrition, disorganization, medical neglect. Yet, for every four or five youngsters that are so maimed, there are ten or twelve that are not. But the insensitivity of the educational system (and the educators) tars them all with the same brush. Some children are slow and unstable; ergo, all who come from the same area are slow and/or unstable!

A low- to moderate-income housing development was built near one suburban elementary school. The children from the development were predominantly black, while the rest of the student body was white. The teachers complained greatly about the new educational problems which had been "dumped" on them—educational retardation, poor behavior, and so forth. One teacher interviewed the other teachers to see whether, in fact, the teachers were having difficulty. Each teacher stated that she was getting along well with the new children, and that they were performing as well as anyone else in the class; it was the *other* teachers who were having trouble: it was *their* children who were doing poorly, making trouble in the lunchroom and playground! Thus are myths born. If some children are not slow, then teachers may act as though they ought to be.[44]

In his introduction to a review of the educational research about the disadvantaged, Gordon makes the point clearly:

> *With rare exceptions the available research relating to the disadvantaged treats the target population as if it were a homogeneous group despite the mounting evidence that heterogeneity within the several subgroups so designated may be a more crucial problem in educational planning.*[45]

WHO SENDS THE MESSAGE?

One author puts the case succinctly, "We are convinced that the great problem [of teaching disadvantaged students] is not the low aspirations of parents and students, but rather those of their teachers."[46] A study of the differential perceptions regarding student achievement showed a clear color difference between black and white teachers.[47] While white teachers felt that their educational failure with black students lay with the students, the black teachers blamed poor facilities and overcrowding. Although both groups agreed on some positive attributes of their black students, white teachers were more apt to see the students as happy-go-lucky and lazy—black teachers viewed them as ambitious. Moore claims that the race of the teacher is less significant than the social class outlook: teachers are middle class, or upward mobile, and no matter what their color is they imbue lower class children with fewer educationally positive attributes.[48]

One recent study of differential perception of learners is that made by Leacock.[49] In this study, the significant findings were the differences in expectation level determined as much by social class as by race of the child. In the lowest socioeconomic black school, the teachers clearly conveyed their low expectations for the students; in the middle income black and white schools, the students were held not only to higher behavior standards, but also to higher learning expectancies. The lower-class black students were not expected to behave very well, and they did not; they were not expected to learn very much, and they did not. The teachers were not punitive nor "bad," but they clearly transmitted by manner and pedagogical style that they did not think these children could learn much. The researchers observed:

> *The teacher working with white middle-income fifth grade children took responsibility for her own limitations and spoke of reacting to lack of attention as a cue to her, indicating the need for her to rouse their interest. On the other hand, the limitations of the teacher working with low-income Negro children were ascribed to the children, and boredom on their part was attributed to their presumably limited attention span.*
>
> *We also observed that a generally friendly teaching style did not prevent a basic nonsupportiveness of learning in the low-income Negro classrooms. The second-grade teacher was warm and motherly and seemed genuinely to like the children; the fifth-grade teacher was friendly and trying to do her best. Yet in both cases the children's very being, their existence, as well as their contributions, were being denied or undermined. Albeit pleasantly, lower status roles were being structured for the children, in contrast with the middle-income white classrooms; poorer images of themselves were being presented to them.*[50]

Of particular interest in the Leacock findings are the reduction of approval on the part of both teacher and student for the students with the best ability and highest IQ. These children, in the low socioeconomic group, were deviant; they were performing better than expected, and thus they were punished by *less* teacher approval, which was also echoed by lower peer approval. It is true, as studies of gifted children have shown,[51] that the very bright tend to be rejected by both teachers and their peers, and are themselves not secure in their sense of self-worth. But at least in middle and upper income levels school achievement does get rewarded.

The very vital role of peer view of achievement is underlined by Pettigrew in evaluating the components which enter into black achievement. In situations where achievement (in white terms) is approved, and there is no implied threat, black subjects did well.[52] The extensive research by Katz and others makes clear that achievement by black subjects is not a simple matter of segregation, desegregation, or integration, but that many factors are present to influence black performance: race of teacher, ratio of blacks to whites, perceived competitiveness, or limitation on rewards.[53]

What is particularly distressing about the Leacock findings is that the mythology of lower-class stupidity is so pervasive that evidence of lack of stupidity is actually resented! These findings correspond with those obtained in the well-publicized study of teachers' expectations made by Rosenthal and Jacobson. In this study, the researchers report that they found teachers reacting unfavorably to children who succeeded, particularly if they were in the slow track. The higher the IQ gain, the greater was teacher disapproval.[54]

Although the Rosenthal and Jacobson findings have been subjected to some severe critical assessment as to the validity and reliability of the research results, the acceptance of the findings by educators has been high, as reflected in the many times the research findings have been reported in the literature, or referred to in connection with other statements or studies. This high acceptance of somewhat disputed findings probably reflects the experience of most educators that teachers' expectations are critical in the everyday classroom experience. It is part of the folklore of the educational profession that teachers who expect bad behavior get it, and those who expect good behavior will elicit such a response from students. Observations of the affects of ability grouping on student attainment also bolster the Rosenthal-Jacobson research. Students assigned to lower tracks not only continue to do poorly, but are more apt to drop out of school, cause discipline problems, and—most significantly—are populated primarily by students from lower socioeconomic groups, boys, and minority group (black or Spanish-speaking) students.[55] The impact of the self-fulfilling prophecy due to being labelled slow or stupid because of lower track assignments is clearly recognized by Kenneth B. Clark in his 1970 proposal to the Washington, D.C. Board of Education. After studying the extensive failure of the schools of Washington to educate students in the basic skills, he recommended an all out concentration on reading and arithmetic–*in heterogeneously grouped classes.*[56] The message being sent to the child by the educational institution may be a major deterrent to his achievement, and this factor may be further contaminated by racism.[57]

SELF-CONCEPT: ANOTHER LOOK

The functioning of the concept of self in regards to achievement has been the subject of study for some years. It was this concept that I attempted to explore in the original statement in *Negro Self-Concept.* Since then, however, a redefinition of the concept seems called for. The simple notion that one is merely responding to how the outside world views one's potential to achieve, as the major ingredient in achievement, ignores other salient forces. Coleman, for example, in reporting his study of the impact of desegregation on school achievement notes that the data are inconclusive regarding many factors which contribute to school achievement. Yet significant among the elements that appear to make the most difference in achievement, is the sense of power that a child may have. If he feels that he is the master of his fate, that he can control to some extent his own destiny, that if he works hard things will go better for him, he is then likely to achieve more than equally endowed students who do not share this feeling. Students who feel themselves victims of a fate in which luck plays the greatest role in who can succeed, believe that it is who you know rather than your own merit which rewards achievement–these children tend not to achieve as well. As the summary of the report states, " . . . for children from disadvantaged groups, achievement . . . appears closely related to what they believe about their environment: whether they believe the environment will respond to reasonable efforts, or whether they believe it is instead merely random or immovable."[58]

The shift in the view of what is known as self-esteem or good self-concept, is exemplified by the work of Smith. He states, in discussing his study of Peace Corps volunteers, that he moved from an earlier notion of "positive mental health" towards one of "explorations in competence." The psychological nature of competence, as Smith reports it, includes factors such as "warranted self-confidence, commitment, energy, responsibility, autonomy, flexibility, and hopeful realism," and also that those who tended to achieve the most, in the Peace Corps situation, showed "inventiveness, initiative, job-elaboration, and responsiveness to challenge . . ."[59]

Positive self-concept, in these terms, is thus viewed as a result of feelings of competence. Competence, in turn, depends on how one comes to view the world around one and one's ability to deal with this world. To feel that one has some power over his own destiny, and can, furthermore, master satisfactorily new challenges that arise, results in achievement *in reality.* Such a view is implicit in the writings of Erikson[60] who identifies levels of development. Each level of growth is defined in terms of behaviors which could be considered achievement of competency *at that time* in the individual's developmental growth. Achieving a sense of autonomy, for instance, is a major task for the infant: he recognizes that he is a separate entity, capable of independent action. A smothering mother, or one who is absent, indifferent, excessively punitive, or inconsistent, may reduce the infant's ability to achieve a sense of independence and this feeling remains effectively instrumental even in adult life.

In his discussion of competence, White makes a crucial point regarding the need for *successful testing* of one's own efforts. As he puts it, the educational setting must allow the child to experience "feelings of efficacy, a growing mastery of whatever he is studying that contributes to his sense of competence. No great educational victory is won by giving a child a feeling that he is loved, but a very great triumph can occur if this feeling releases previous inhibitions so that he becomes able to exert himself, experience efficacy, and start increasing his fund of confidence that he can deal adequately with the educational environment."[61] White presents the importance of the development of a sense of interpersonal competence, which occurs when one's view of what should be done prevails—but only when there is something or someone *significant* to prevail over! There is no great triumph in bullying one's younger siblings or weaker peers on the block; but there is a sense of power in compelling a school principal to introduce a new program of black studies. As White says, "It might be difficult to convince a social planner that the best way to evoke a sense of competence in those for whom the plans are made, would be to let them change the plans over his stubborn resistence. Yet if this actually happened he would be showing extraordinary talent for his work"

The crippling effects of rigid structure over which the child has no control are noted by Beker et al.[62] "The chaos at Downtown school when the teacher leaves the room seems highly significant. . . . Accumulating evidence supports the notion that people who are trained to function with tight external control tend to become dependent on it for effective behavior. . . . This . . . points up a serious dilemma. The assertion is frequently made, particularly by inner-city educators, that disadvantaged children need tightly structured programs . . . Taken alone, however, this approach tends to perpetuate the very dependence on external control that may be a major component of a possible 'disadvantaged syndrome.' "

Beker continues:

> *It is relevant, although not new, to point out that even the rebelliousness of these children may be an expression of their dependence on external attention and limits and of their lack of developing identity—as reflected in the inability of some of them to recognize their own handwriting and words. These tend to be low-status children, and they know it; and they will be helped little to feel better about themselves, or to achieve, by continued suppression, although this may be the only way in which the school as now constituted can deal with them at all.*

The concept of the self-actualizing person, who moves from a feeling of competence and control in some areas of his life into further extensions in his own behavior, and the ways in which he allows others to become self-actualizing, are at the core of many modern educational programs.[63] Gardner,[64] in talking about self-renewal, bases his argument on the value of continued response to new challenges, and that every such successful response produces a wish for more such deeply satisfying experiences: the person then is never alienated psychologically or emotionally passive or defeated. What then about the Black Power movement? Does this have some message regarding the concept of competence (positive self-esteem or self-concept, if you will)? Thomas makes some pertinent points when he states:

> *The literature on motivation and human needs, with some slight modification, also can be generalized to the black*

> *movement. Self-esteem and self-actualization needs are illustrative. Independence and a sense of adequacy in dealing with one's own problems are indices of high motivation and optional [sic] adjustment. These needs, when fulfilled, permit one to experience recognition and respect, the right to make decisions about his own destiny, and to have a valued role in society.*[65]

Thomas concludes his analysis of the role of black unity and black-white confrontations as developing the critically important view, by blacks, that, "Like children who have grown up, black men know they are boys no more."

One must be cautious, however, in giving too much credit to the feeling of competence as providing all that is needed for achievement in school or out. As Birch and Veroff state, "Although competence and the motive to achieve are probably highly related theoretically, they are not the same thing."[66] For a sense of competence to *produce* achievement in reality, it is necessary that the individual (or the group) have a realistic sense about the ultimate effectiveness, or utility, of their demonstration of competence—in other words, to achieve. Women, for instance, may be aware of their competence in academic achievement, but in numerous instances do not demonstrate such competence because to do so would scare off potential mates. Competence in this sense, then, has a negative utility.

Another example is the struggle of the black cooperatives in the South. Here one finds a strong motivation to achieve, and the farmers who are working together to farm the cooperative have demonstrated their competence in farming. Yet the institutional framework within which they must function, an economy effectively controlled by whites who see cooperative enterprises by blacks as a threat, has made the number of cooperative ventures very few indeed. Many have failed. The struggle to survive is paved with harrassment, intimidation, interference with getting produce to the market. How much *visible* thwarting of the motive to achieve, based on earned and learned competence, can be experienced before an individual will respond by giving up? And what will his children have learned from such failures in being able to demonstrate competence? As Birch and Veroff point out, groups differ in their experience with social frustration. Jews, who are typically considered (and consider themselves) to be highly achieving, end up rather differently when contrasted to another immigrant, non-English speaking group, the Italian-Catholics. What cultural norms supported Jewish upward mobility, and tended to confirm Italian acceptance of a lower socioeconomic status quo?[67] Again, the perception of eventual efficacy in achieving perceived goals, as well as the desire to attain such goals, seems crucial. Here again the Black Power movement may well serve as an intervening variable: for the first time the black citizen can realistically perceive that he may have a chance of realizing the goals he has been motivated to want, which in turn induces a sense of increased competence in an ever widening circle of social (power) relationships.

WHO BENEFITS FROM THE BLACK REVOLUTION?

The most obvious beneficiary of the black revolution is the educated middle-class black. Already endowed with skills and education, he can step forward into positions of prestige and related higher income. His children can aspire to good private schools (and expect more favored treatment in admissions policies anywhere) and on to better so-called white colleges. Thus, those who already have, get more.

This newly mobile group, moved now into positions of leadership visible in both the white and black communities are, in many instances, only one generation removed from rural or urban poverty. Many of the new leadership elite have had a full experience of the "no dogs or colored allowed" policies of Jim Crow America. Such positions do not, of themselves, endow an individual with feelings of safety or of merit rewarded. The new prominence, the titles, the front desk, may be only the window dressing needed to show how "integrated" a company or college has become. One black intellectual stated the perception of his new eminence in university affairs: "I wonder if I am placed on a powerful committee so that there will be one black face, or is it because I am really competent?"

In responding to the black revolutionaries' demands for instant power, institutions rush to show how "integrated" they are. The result is, as noted above, many blacks placed in strategic positions, and the election of a few to political posts traditionally held by whites. Although the elections of Stokes as Mayor of Cleveland, Hatcher as Mayor of Gary, or Gibson as Mayor of Newark seem to be a recognition of quality of leadership, one cynical black observer said, "Sure, blacks get elected mayors of cities when the city is so bankrupt nothing can save it. Then when the city collapses in a morass of financial tangles and civic deterioration, it can neatly be blamed on the incompetent black leadership." Time may or may not support this prophecy. What is critical, though, is that the belief (even by those who benefit most from the black revolution) that such benefits are transitory and motivated by white self-interest rather than recognition of black competency, may produce in time the very effect such white strategy was employed to abort: the birth of a politically radicalized middle and upper class black group. Having once tasted the benefits of power, leadership, and affluence, yet at the same time viewing the whole process with some detached suspicion, this key group of black talent may become aggressively independent of their benevolent white protectors. At this point, the black revolution may come to maturity, and the power confrontation will be for stakes far more threatening to the white power structure than any riot or black militant demands.

While this leadership evolution may indeed take place, it is important to see the black revolution from another perspective. The dynamics of change may take a rather different course. Instead of becoming the new leadership for black revolution, the middle class black, moved into new positions of status, may, in fact, become (or remain) a bulwark of conservatism and caution—particularly in black affairs.

The customs of years, indeed of centuries, may not be as easily cast aside as the young black revolutionaries would wish. The middle-aged, middle-echelon black executive or professor has himself been educated to believe strongly that the goal of Negro efforts led into the mainstream of white America. There were the wealthy Negro separatists who developed their own country club life and own jet set (sic), while others became "white Negroes" (oreos).[68] These upper class Negroes had little to do with other Negroes unless they happened to be in the same professional or business circle, nor did they seek out whites or other Negroes outside their normal working life. But whether separatist or amalgamated, the Negro of the 1940s at least perceived that school desegregation, and desegregation in the armed forces, in unions, in industry, in housing, was a major goal. The doors of choice were to be opened—the black could become just another neighbor on the block or choose to stay within his own ethnic island. The goal was true freedom of choice.

The generation which is in its mid forties today may find it hard to decide whether its allegiance is with the black power brigade or the integration-oriented Urban League and NAACP. It is the offspring of the mid-forty generation who are pushing the confrontation and demanding a decision.

The bastion of Negro middle-class ambitions, *Ebony* magazine, was picketed on December 30, 1969, by a group of young black activists who were asking that the magazine reorient its position toward a more militant, more clearly "black" position. The black who is an established professional, or who has newly moved into a position of relative power and status, is caught: his own children come home wearing Afro hairdos and dashikis and play soul music at top volume on the expensive stereo tapedeck. They seem alien and even threatening. The middle class contemporary generation gap is critically apparent in many middle class black homes: the parents have "made it" into the safety of civil service or academic life, and their youngsters are now so sure of their position, or of the security of having an adequate family income and family protection, that they can afford to flaunt authorities at Ivy League colleges.[69] Their parents, to whom Harvard or Cornell were impossible dreams, are affronted and afraid when sons and daughters at prestigious institutions seem to be throwing away, through defiance of rules and through black power attacks of all kinds, the very type of opportunity they (the parents) could not even have considered, but valued highly.

This conflict of generations is still too young, too fragile in the context of endemic interracial violence, to predict future responses of either the older or younger generation. The position taken by the older black man or woman, now in jobs of status and prestige in the white world, is uncertain: Will he turn increasingly towards the separatism of the Black Power movement, with possible jeopardy of his job security in the gamble for more significant exercise of power? Or will he take a low keyed approach to race relations, with the risk of alienating his sons and daughters, to forge stronger ties and links with white power groups?

The Black Power movement has been seen by many as a youth movement: it is the younger people who are the most strident, attracting the greatest publicity with uncompromising demands accompanied by clear threats of violence. But are all black youth part of this movement, either actively or by passive-but-supporting approval? The lower class black youth now entering high school or the world of work are primarily urban born and raised.[70] They are linked to the rural past by their parents' feeling of powerlessness, hopelessness, and the struggle against overwhelming poverty. But they are urban, and as such are more exposed to city ways, city television, and—despite claims of critics—the wider education available in the city schools. The city school may be bad, but rural segregated education is and has been infinitely worse.[71] In this environment, are black youth turning more militant, or more aggressively motivated toward the kinds of middle class lives other second generation youth have sought? An interesting light on the goals of urban youth was cast by the first troubled year of Federal City College, the first Federally supported college for the District of Columbia. In 1968 the college opened with such a heavy demand for admission that entrants had to be selected by lot. Furthermore, the faculty was caught in a battle over whether the college should focus on black studies for a black community, or offer the typical standard fare of a four year city college. The students elected the latter. To the surprise of some of the faculty who were insistent that at least half the resources of the college be devoted to black studies oriented in the main toward revolutionary goals, the students opted overwhelmingly for bread-and-butter courses—courses that could get them into upper grade government jobs and on to industry or graduate school if desired. Not as many as had been expected rallied to the cry for black relevance. The opening of the college in September 1970, as reported in the *Washington Post,* has radicalized many who might have been characterized as the "bread-and-butter" group above. Because of Congressional lack of interest or deliberate policy, the college opened unable to take care of several thousand students who had been given unqualified admission. The resulting community outcry, and the rage of the disappointed students should not be underestimated. This kind of official action (whether accidental or deliberate) is fuel for the black revolutionaries, and also for the antiblack hate groups who relish each manifestation of black violence: "See, they are still savages."

Conflict is growing between the spokesmen for the black revolutionary concept, such as Nathan Hare,[72] and those who believe that black separatism is the wrong approach, such as Andrew Brimmer of the Federal Reserve Board.[73] The former position calls for black studies which include every standard discipline, and all taught with an eye towards developing black leaders to work in the vanguard of the black revolution—whose goal is the taking over of the sources of control of a significant segment of society, if not the overthrowing of the total social structure eventually. The latter position, exemplified by Brimmer in his critical position on the FRB (sometimes called the fourth branch of government because of its independence of either presidential or congressional influence), states that for blacks to put their energy into efforts to build black capitalism may be a mistake. He is quoted as saying, that "black capitalism may retard the Negro's advancement by discouraging many from full participation in the national economy. . . ." Bayard Rustin, whom many militant young blacks consider an Uncle Tom, does not support the notion of black studies as *the* major pathway to the realization of the goals of the Black Power movement.[74]

In a survey of the black graduates of Spring, 1970, *Ebony* reports that most black college graduates want jobs where the money is. Some will use their skills and advanced training to return to the black ghetto or to the Southern black belt. Most black graduates seem to have the same "security-jobs-money" orientation as their white peers.[75]

The split between activist black youth and the more cautious older generation is exemplified by studies of those who participated in the big city riots. For instance, in Watts, persons who were born in California were more apt to favor the disturbance and to be participants than those who had come to the area from other states. Those who participated in the Detroit and Newark riots were more likely to have been born and raised in the North; in Detroit, six out of ten rioters were born in Detroit, while seven out of ten *non*-rioters were migrants from outside Detroit. The Newark data are particularly interesting since over half of the non-rioters were born and raised in the South, while three-fourths of the rioters were non-Southern in origin. The pattern of arrests also showed that Northern-born rioters were apt to be arrested on the first day, indicating more propensity to be the first to "engage the enemy." An analysis of the meaning of *Black Power* to the Detroit black community in 1968 found that the slogan was more apt to be favorably viewed in a militant context by Northern-born blacks; Southern-born blacks saw the slogan as less favorable, and were less likely to see the concept as a militant one.[76]

The direction that the black revolution will take in the years to come is difficult to forecast. The assassination of Martin Luther King, Jr. with the aftermath of rioting and destruction, has not only forced new assessments of the black situation, but it also left a leadership void of a particularly critical kind. Dr. King was able to rally significant political support from the white community, and he could also provide a sense of power and self-determination to the lowest strata of the black community, all of the time working at the vulnerable points in the system. He made the system work *for* him, and also could identify the source of violence as residing in the white racist reaction to his powerful pressure. With his death, no similar leader has emerged to use the system so effectively for the amelioration of the black condition. What has emerged is the leadership of the Black Panthers whose rhetoric derives from spokesmen as dissimilar as Eldridge Cleaver and Frantz Fanon.[77]

The role of the Black Panthers in gaining not only black militant but also white liberal support is instructive. The response of the president of Yale University to the intense interest of students in the trial of black youth in New Haven in May 1970, must not be underestimated. Extreme as some of the leaders appear to be, they and their confrontations with the law have been dramatically able to focus public attention upon discriminatory justice which has long forced blacks into frightened if not servile compliance.[78] It may be a long time before justice becomes color-blind, but the contemporary legal battles of black militants are teaching blacks in significant numbers that they, too, can and should demand equal justice under law.

The events of May of 1970 make the task of prediction difficult. While the country as a whole was convulsed with grief, anger, and outrage at the Cambodian war strategy, and the related shooting of four students at Kent State University in Ohio, the black community was aware that no such national outcry occurred when students were killed and injured at Orangeburg, South Carolina a year earlier.[79] The wanton shooting at Jackson State College in Mississippi, however, a few days after the Kent tragedy, reminded observers that there was a differential response to white and black death, though belated governmental interest attempted to head off another series of black reprisals.

The piling up of these events highlights another aspect of the problem which makes predictions about the future difficult: the press no longer completely ignores outrages perpetrated on the black community. There is apt to be some public dismay at the continued utility of white racism in the political campaigns of Northern as well as Southern leaders.

BLACK POWER AND BLACK SELF-CONCEPT

A reassessment of the role of the range of actors involved in the development of any individual's sense of a competent self leads us to examine with more care who sends what messages to whom. The child who learns that he is not capable of learning may be the victim of the distortion of the perception of others. The significance of the Black Power movement is therefore not only to enable black persons to see themselves more accurately, but also to break the stereotyped view which the world at large has imposed upon them.

The folklore regarding the black population was that they were more impulsive, more prone to violence, less inhibited. Yet the facts do not prove this to be necessarily true. What was true was a differential application of law which condoned criminal acts when undertaken by blacks against blacks, "because that is what they are like." For example, at Hunters Point, a predominantly black area of San Francisco, and scene of a disastrous riot in 1966,

> *The police do not regard the kind of boisterous behavior frequently complained about by tenants (noisy parties, gambling in the halls and stairways, and the like) as police matters. They feel that "this is not an upper-class neighborhood," a judgment based on the common sense observation that behavior in this neighborhood is different than it is in others, and that it would only harass the population unnecessarily (and make police work harder) to police it strictly. Sophisticated as this view may be, it implies to many residents a paternalistic and racist attitude.*[80]

While some individuals were angered that police would tolerate such behavior, they would also be infuriated by police attempts to impose "middle-class values" with the implied prejudices of such efforts.

Yet the example of differential law enforcement standards is echoed by other ways in which those already marginal to the system are further defeated by it. All services are less apt to be invoked for the poor than they are for more well to do. Park maintenance, garbage collection, school financial support, are notoriously less adequate in the less adequate parts of the community.[81] Because of curious procedure in figuring out state support for handicapped children, Montgomery County, the richest county per capita in Maryland, would seem to have more children with handicaps (and thus eligible for state support) than the city of Baltimore, where there is a high incidence of handicapped children.[82] The self-fulfilling prophecy that the poor are not worth much and therefore should not get much makes the poor much less able to become not poor.

Breaking the cycle of such forces has become the object of governmental policies, presumably, and nongovernmental pressure groups as well. The black community has been radically stirred by events since 1954. Many black leaders in the late 1960s were outspoken in their opposition to white cooperative efforts, identified white values as basically racist and materialistic, and increased their emphasis upon black solidarity and positive black self-acceptance. How have these events and pronouncements affected black youth? Black spokesmen state with assurance that the movement for black identity and Black Power has had a significant impact upon black Americans. The proliferation of black-oriented magazines and journals attest to a belief in an interested, primarily black, audience.[83] The development of black studies programs on college campuses across the country[84] is a visible demonstration of the claims that Black Power is a valid social movement deserving academic recognition.

Reports in the mass media[85] and personal encounters seem to provide a safe basis from which to generalize about a *qualitatively significant change in black behavior.* Black professionals readily speak up in most mixed groups; blacks are reported to be asserting themselves in encounters with public officials—police, welfare workers, school personnel, government representatives, and armed services personnel. Black youth in public schools in many parts of the country have led strikes and sit-ins and presented demands to school officials ranging from a ban on playing Dixie to soul food in the cafeteria, to firing racist teachers and including black studies in the curriculum. These events cannot be discounted. The pervasive element, particularly striking when one has experienced the Uncle Tom responses typical of pre-1954 Negroes, is that *today's black individuals exercise personal assertiveness because there is no longer an expectation of crippling retaliation.* Students who protest do not expect that they will be suspended or have their grades lowered, or if these things happen, they swiftly call upon legal resources and publicity that were previously unavailable or unsympathetic. Every institution in America—churches, schools, unions, and local, state and Federal units—all have felt this pressure and have not remained uninfluenced. Efforts to suppress the new black leadership are increasingly unsuccessful, even in Mississippi.

Since our definition of positive self-concept includes the concept of competence, which in turn subsumes feelings of power, then there is ample anecdotal evidence that black self-concept has been significantly affected by recent events. Baughman, in presenting an analysis of recent research, argues that too many studies of blacks have dwelt with their presumed negative self-esteem. As he states it: "The literature on the black has, we believe, given undue emphasis to his subservient behavior and roles. (We frequently forget that countless whites act in a similar fashion.) Too little attention has been devoted to blacks who evidenced pride and competence. . . ."[86] Baughman advances the thesis that since most black children learn about themselves within a supportive black family and black world, they gain a sense of solid self-worth. It is only when they contact the white world, which may present another view, or which forces upon them evidence of peer achievement compared to whites, that their positive sense of self is threatened. These youngsters may then revalue themselves downward, or retain a positive self-concept along with a cynicism and suspicion of the white world.[87]

At this point the discussion of self-concept needs additional clarification. As Brookover and Erickson make clear, "that although a significant proportion of students with high self-concepts of ability achieved at a relatively lower level . . . practically none of the students with low self-concepts of ability achieved at a high level."[88] One can have a positive *general* self-concept, and yet have a poor image of oneself as a *learner.* It is very possible, indeed it might be essential to survival, that an individual who finds achievement impossible or blocked in one realm of life find other rewarding areas for success, and the concomitant feeling of enhanced self-esteem. If a black male succeeds less well in school than black females, or all white students (which is the statistically typical case), he can find his support in the approval of the street gang. He can win peer acclaim through daring assaults against authority. A study of successful and unsuccessful persons from seriously disadvantaged backgrounds undertaken for the Department of Labor Special Manpower Programs (1970), provides interesting evidence: Blacks and Mexican-Americans who "succeeded" were apt to have identified—even in the ghetto—with what the authors call mainstream life. This is characterized

> *by a work achievement ethic, close family ties and loyalties, avoidance of trouble with the law, stability on the job, taking responsibility for one's own destiny, orderly planning for long-range goals, the ability to sustain activity in goal-directed behavior, and an ability to make somewhat harmonious adjustment to the existing [white] social order.*[89]

By contrast, unsuccessful individuals were more apt to have participated actively in "street life" or gang life which was characterized by

> *an ethic of toughness, shrewdness, hustling, violence, an emphasis on having a reputation among one's peers, a lessened concern with family responsibility and ties than with ties to peers, a glorification of antisocial acts (in the mainstream sense), and the absence of long-range planning or goal seeking, in favor of immediate gratification, all combined with a rejection of mainstream values.*[90]

Those who identified with mainstream values also had, interestingly enough, positive recollections of school officials and school life, while those who identified with the gang or street life style had negative memories of school, reported conflict with school authorities, and, if they were black, had dropped out of school.

A boy who is rejected by the street gang can either bolster his self-esteem by modeling his behavior after those whom the gang approves, and thus winning his way back in, or he can find other models—parents, teachers, group leaders—who offer sufficient rewards to make up for peer disapproval. A researcher who finds high self-esteem in a group which has disavowed the more obvious values of the world around them should not be surprised: such individuals construct (or find) others who can feed the need for self-esteem which is essential to personal survival. A person convinced that no one finds anything to like about him has no recourse but insanity, suicide, or flight through drugs or alcohol.

One of the most powerful effects of the Black Power revolution has been to make mainstream life rewards within the grasp of the average black individual. Whether the black revolutionary seeks black-instigated apartheid, or merely a more powerful voice in black-related affairs, the fundamental goals are not too different from what one might consider mainstream American life goals: feelings of pride in self and group, security for self and family, self-enhancing work and leisure activities, freedom from hunger, violence, illness, and control over one's own destiny. Thus, although the black power advocate appears to reject white society's assumed values, he is not in actuality very far removed from his upward-striving, nonrevolutionary black brother.

An interesting example of the way in which black pride is disassociated from white values, even though the underlying objective is congruent with mainstream American life, is a discussion by Staples of marital role training needed by black males. Staples points out that marital stability requires a reorientation of the concepts of adequacy which are typically exhibited by lower-class and impoverished black males. This can be accomplished, he states, by "inculcation of pride in their African heritage, by a reminder of their forefathers' patrilineal and patriarchal forms of family organization." He goes on to say:

> *The use of the term patriarchy may be somewhat misleading to some readers. This does not imply a regression of [sic] the equalitarian family model now emerging in certain segments of the black community, most notably among the middle-class strata, but rather means a greater male participation in family activities, including economic support and socialization processes. If necessary, however, male dominance as a consequence of the psychological identification of black males with African family systems is not excluded*

> *as a viable substitute for the female-dominated households that now prevail in certain segments of the black community. There are few objections to this idea in the black community, especially from black females.*[91]

By invoking an African patriarchal system the author appears to reject the basically patriarchal mode of mainstream American life, women's liberation to the contrary. Rather than saying, "let's have old-fashioned American families where the father is the head of the household," he must invoke an African pattern in order to make the stable family with male dominance an acceptable concept for blacks to strive for.

Whatever the rationale for changed behavior, educators must be concerned with youth—black or any other color—who have sought to change their life style, for whatever reason, into one that conforms to mainstream America, but find little if any support for these efforts. The rather bitter argument over the perpetuation of the ghetto dialect as a distinctive linguistic form illustrates the dilemma.[92] Defenders of black dialect see it as a valued cultural expression; critics point out that persons who speak (and write) using black dialect forms will be excluded from all but separate black communities and enterprises. The pragmatic black advocate knows that getting a slice of the economic pie or political power in this society depends on some of the outer attributes of entry—speech, dress, manner, skills. Once access is gained, the uses of economic or political power can indeed be turned to benefit the black community. But until then, ghetto life style and ghetto speech are an impediment.

The tragedy of the American Negro of past decades is that he exemplified in many significant respects the core values of mainstream American society—piety, family loyalty, compassion, hard-work ethic—only to find that living by these virtues did him little good. He still suffered gross injustice, insecurity, and wilfull violence. White society—until the current Black Power movement—made it clear that no amount of "being good" could obscure a black skin.

The conflict which still traps the black individual is therefore one of finding assurance in a white and racist society whose life style he finds agreeable and values, but which reluctantly, capriciously, and minimally provides assurance for black achievement on a color-blind basis. When a black colleague was promoted from assistant professor to associate professor in a major university (one with very few black staff members), he could not help knowing that some of those around him felt his promotion was only because he was black. The fact that he had published widely and well, actively participated in leadership roles in professional organizations, and was also an effective and conscientious instructor was irrelevant. And, because of these doubts, can he himself be completely sure that his promotion was based purely on merit on a competitive basis? Was it a cynical ploy by the university's administration to forestall black student and staff criticism on the basis of their previous racist history? What if he had been white? Would the promotion have come as soon?

If these questions haunt a sophisticated academician, and at some cost to his own sense of self-esteem, what then of the vulnerable and naive child? The messages he gets are still, in the 1970s, laced with doubt, cynicism, and hope at the same time. Researchers report that when black youth marry, skin color of spouse remains a significant factor (as of 1965), affirming data which was documented in the 1940s.[93] A black male will consistently prefer a lighter female, and this choice has not changed over several decades. Despite the traditional disadvantage associated with dark skin, the 1965 researchers reported that status-mobility orientation was *less* with lighter skinned men than darker. As the researchers comment:

> *... we clearly underestimated the impact of the Negro struggle for racial equality on the skin-color attitudes and self-concepts of men in the Negro community [the District of Columbia] surveyed, and therefore did not anticipate that a reversal of traditional findings could already have occurred. Our data strongly suggest that the trends in this new direction started at least as early as the late fifties and are therefore associated with the entire period of Negro activism. The current popular glorification of blackness and Negro characteristics generally appears in perspective as the continuation of this trend . . . at least for men.*[94]

As Udry and his associates note, skin color and mobility striving was only a source of frustration for the darker men who married prior to

1960, but "If this motivation [to achieve] comes before a man has crossed all his status bridges, it can be converted to a status asset." Another study, also drawing on data collected during 1965 and focussing on skin color, mobility, and attitudes towards whites, notes that:

> *Taking account of the earlier findings on the color barrier, a cumulative pattern can be noted: dark color and feelings of powerlessness interact to produce the lowest returns in income and occupational status and to produce intense hostility towards white society.*[95]

Ransford concludes that only as skin color becomes irrelevant will there be a diminution of violent attacks on whites by blacks who perceive themselves to have been blocked from achievement because of color. Since the research from which these conclusions were drawn was conducted in 1965, Ransford amended his conclusions in 1970 when the analysis was published, with the following interesting observation:

> *. . . there are many indications that color has become less important in the Negro community . . . In fact, for some segments of the black community (e.g., young college students), the traditional evaluations appear to be reversed—dark skin is now admired and light skin is not. A provocative question is whether the black pride movement has been powerful enough to override completely the traditional stigma of dark color. It seems unlikely that this has occurred for a cross section of the Negro community. One can speculate that dark color will not lose its negative evaluation until a new generation of black children has been highly exposed to "black is beautiful" values and has replaced the current generation in positions of power and status. Further, although a dark color may have lost some of its negative aspects in the black community, very likely white society is still allowing higher proportions of lighter and more Caucasian-appearing Negroes into better socioeconomic positions.*[96]

In interesting support of Ransford's conclusions is an analysis made of several opinion polls conducted between 1964 and 1969. The data reveal rather clearly, according to the author, that age and education are major factors in the attitudes of blacks and whites, with education being the more significant. For instance, "college-educated white youths are more sympathetic to protest methods than less-educated white youth; but black youth, and black college youth in particular, are readier to embrace protest methods than white college youth."[97] Although younger blacks were more apt to advocate violence than older ones, and college-educated blacks more than non-college, it was also noted that:

> *education moderated anti-white perception among young, better educated Negroes.* Less *educated black youth evidenced the most extreme forms of hostility against whites. Enlarging the black* college *population, then, could be expected to lead in the long run to improved perceptions of whites by blacks, but not necessarily more peaceful relations with them.*[98]

What do these various studies indicate? As Goffman, in his classic discussion of the ways individuals cope with stigma points out, individuals who carry such stigma, whether on their bodies, on their records, or their souls, must not only learn how to cope with the stigma themselves, but also how to help others around them cope.[99] The avenues open for this dual responsibility are neither clear nor smooth. The concept of self, in which a recognition of stigma must be encapsulated, evolves through stages of self-disgust, despair, resignation, acceptance, and, if the individual is particularly fortunate, achievement that neutralizes the stigma completely so that it becomes not irrelevant, but rather a source of inner strength because of the odds that have been overcome.

The Black Power movement then, has focused direct attention upon the source of the stigma which had debased and destroyed generations of blacks, and by stating that the stigma of color is to be valued, and that the problem of color is a white one, not a black one, effective coping mechanisms are brought into play. The response to this changed view of self, in which the schools became a major source for providing data on achievement and access to further achievement, is no longer one for ivory tower speculation. If the current groups holding power over school programs do not respond appropriately to black expectations then a confrontation is inevitable.

WHAT CAN THE SCHOOLS DO?

The role of the school in the cycle of black despair described by social commentators and reported by researchers[100] is one of benign neglect. Like the police, in their response to black behavior at Hunters Point previously cited, school personnel tend to behave as though they knew that most slum dwellers were, in fact, less able to perform well. Educators have for too long assumed that poor people and/or blacks have a different value system from that held by the predominantly white (or black) middle-class society. This is a persistent myth. Nearly two decades ago Havighurst and Taba pointed out that middle-class and lower-class white youth responded similarly to questions about right and wrong behavior. The message that the culture had conveyed as to what one was supposed to believe and to do had been well learned.[101] Current research shows, again, that inner-city black youngsters and adults do not, in fact, reject so-called middle-class values.[102] As noted previously, the impact of a minority of visible children with difficult educational problems obscures teacher perception of the majority of the children who *could be* similar to any heterogeneous group in any classroom.

The solution appears to be perfectly obvious. The educational system must be so equipped that it can, *in fact,* treat each child as an individual and a contributing member of a social-peer group.[103] The mass-class-group procedures utilized in almost every school must be eliminated. Every study of ability grouping has shown that differences among members of any class so grouped was greater than that between adjacent groups.[104] No score or scale has yet been devised to use in selecting truly homogeneous groups for instructional purposes.[105]

It is perfectly clear, too, that the key adults in the decision making about children and youth need extensive reeducation. The kinds of observational studies conducted by Leacock,[106] Smith and

Geoffery,[107] Eddy,[108] and Fuchs,[109] should be replicated in the classrooms of every system, every year. The observations and critiques by Herndon, Holt, Kohl, Kozol, and Decker,[110] are often apt to tell us more about the writer than about the conditions of schooling which produce the appalling pictures of educational destruction in city schools. These personal commentaries are useful; they have at least reached the general reading public in ways in which educational and social-psychological research does not. But they also distort the picture as well as provide somewhat dubious, sentimental versions of how the horrible conditions they describe can be eliminated.

Teachers should have feedback on their differential perception of student ability. As Spindler pointed out some years ago,[111] the most well-meaning and presumably adequate and sensitive teacher may unknowingly distribute his rewards—and thus make learning more possible—in terms of his perception of the socioeconomic level of the child, than that child's true potential.[112] Similarly, the ways in which teachers reward and punish on a sex bias must be reported back to them.[113]

Educators need to take a critical look at some of the assumptions made about who does or does not succeed within the institutional framework. Bowman,[114] for example, points out the well-known phenomenon that school dropouts show a poor record of participation in extracurricular activities. He fails to note that this poor record may reflect inappropriate educational practices, rather than lack of interest. The typical secondary school bars students from participation in extracurricular activities, including highly visible and rewarding interscholastic sports, unless they have the "right" grade point average, no disciplinary marks against them, have the proper gym equipment, can pay fees for uniforms (and cleaning them) or instruments, and are free after school (have no jobs and not needed at home) for rehearsals, practice, or trips. The lower-class youth can rarely qualify. Bowman suggests some educational changes, but he does not suggest that it is the total school structure which must be examined. His myopia is all too typical of educators, locked into the institutional framework so completely that they no longer can observe how it operates. Tinkering with a program here and there—more interesting content, smaller classes, black studies, parent contacts—will not do the job.

A school system based on a class bias, with a built-in failure structure, makes it inevitable that those already victims of the system will find the school more of the same. It has been noted that when observers are present, teachers tend to behave better;[115] it is possible that by using this simple notion, some kinds of classroom procedures could be improved. Techniques which provide teachers with information on their responses to children might then be used. Methods now available must be applied so that no child is forgotten; no pupil could advance through the grades and remain a functional illiterate—which is now the case for thousands of inner-city children and the neglected and impoverished areas of rural America. As Clark so succinctly puts it:

> *The ghetto child needs to sense from his teachers that they respect him as a person. And the only way he can sense it is through his accomplishments and through teachers providing the parameters for accomplishment. This is where the whole thing about pride comes in. I am convinced that black children or any other group of children can't develop pride by just saying they have it, by singing a song about it, or by saying I'm black and beautiful or I'm white and superior. These approaches are senseless. Pride comes through achievement and demonstrable achievement. The people who know this best are the people in the ghettos. The children know when they are able to accomplish something and when they are failing. They know when they are being relegated to the dung heap of academia. They know when they can't read as well as others or when they can't do arithmetic. To set them apart and give their groups the names of birds or animals doesn't fool them a bit. I don't know a single child who is so unintelligent as not to know when his school has given up on him. I don't know of a single child who can be fooled by being told that it doesn't matter whether he knows how to read or write or do arithmetic because he has a glorious culture or because of his great color. They don't buy it. That's why they become junkies, to escape from themselves, to escape from that second-class kind of reality. If you want to change that reality, you've got to change it in the schools. We have to start teaching these children.*[116]

SUMMARY

We started out by examining the possible impact upon black youth of the new emphasis on Black Power and the new black militance. It has been suggested that those who already are able to work within the system, the upward mobile and middle class black will benefit the most, but that the poor and the defeated will remain trapped—and so will their children. The trap, however, is not the product of individual lack of self-esteem, but rather the insidious impact of stereotyping.

The factual realities of impairment due to malnutrition, medical neglect, and ineffectual and inadequate community services for the disorganized and underemployed, which do produce educational problems for many children, have also been utilized by the educational system to entrap thousands more in a failure syndrome. The reasoning goes something like this: "Joe is black; his mother suffered anemia before his birth; he shows the effects of inadequate nutrition and care. Bob is also black. None of these things happened to him. But both Joe and Bob are in the same class in school. They live in the same apartment house. Therefore it is probable that Bob is more like Joe than he is unlike Joe. We shall therefore assume Bob is like Joe. Since Joe cannot learn very well, Bob probably cannot either. Ergo: Bob becomes like Joe—a slow learner." But since Bob is not Joe, and Bob is quick and able and alert, the impact of being told he is like Joe is disorienting and disabling. Bob wonders if, indeed, he may not be like Joe. Yet he knows he can outwit Joe anytime. Bob's sense of his own competence becomes blurred. He gets confusing messages, to put it mildly. Joe may drift into adulthood, and, being not too smart, end up placidly if sporadically in a menial job, if employed at all. Bob, however, uses his submerged and uneducated ability to try to "make it" on the street. Not having learned how to focus his attention, or use his abilities in a disciplined manner, he soon is on the conveyer belt: crime-jail-probation-crime-jail. A rare Claude Brown escapes.[117]

The problem then, is not as simple as merely intervening in ways which may seem to have an effect on the child's self-concept. The need is for clarification of how a *sense of competence is acquired, and how motivation towards achievement is sustained.*

The Black Power movement and black militants are stirring

those whose apathy was only skin deep, ready to respond to the challenge of becoming something they always knew they could be had they the nerve to oppose the system. Since the system is highly vulnerable (the American dilemma in a new dimension), the Black Power rhetoric will indeed work. For those mired in the system only the deliberate actions of alert and educated personnel can break the cycle of personal and social destruction.[118] It will take more than an integrated *Sesame Street* or biographies of Martin Luther King, Jr., to overcome the disasters which occur daily in the average inner-city black classroom. Here is where the central problem for the educator resides. It is the pathology of these schools which needs immediate attention. Only then will we be able to diagnose adequately, and educate adequately, *each* child.

AFTERWORD

It is not within the scope of this essay to develop in depth any prescriptions for education which will make education available to all equally. However, having posed the problem, one is tempted to suggest some solutions, based on experience, observation, gleanings from research, and theory:

1. The problem of black identity is as much a problem for the white student as for the black. No adequate educational program for black children will succeed unless white children are helped to understand racism along with black children. Ideally this kind of education should occur in schools with as wide an ethnic mix as possible.

2. Achievement for the educationally crippled can be drastically improved by

   establishing the legitimacy of teacher accountability.

   bringing in many more adults at every educational level, many of whom may well be from the school community.

   utilizing cross-age tutoring, grouping, gaming, and playing. Older children are superb teachers; no one learns more than a teacher about what he is teaching.

understanding that no child should learn to read a textbook: he should learn to read books, faces, music, moods, pictures. Commercial television can be as useful in any classroom as in any home, though the lessons may be different—but this is the real world.

using the new technology, such as videotaped lessons, to help every teacher see his own teaching, and every child see his own learning—every year, or every month, or oftener.

opening up the school day. Lengthen it, lengthen the school year also so that ample *time* is available for the leisure to learn.

3. In every school system there are some exceptionally successful teachers. They can make the unruly happy, the antic dance, and the mute talk. These teachers (and maybe whole schools) should be identified and studied. What is it that they do that other teachers do not do? Why are they so successful? With what groups are they especially successful? In what way can they become models for other teachers?

4. Administrative personnel must

make it clear that a quiet school is a bad school; only when students talk more than teachers will learning take place.

remove as much entangling paper work from the classroom.

support teachers who are different if their differences are factually productive of learning and love of learning.

create community centers out of schools; open them up evenings, weekends, holidays, so that the resources of the school are available to adults. Bring in new resources—well-baby clinics, health screening clinics, walk-in mental health services, fun and games, classes for adults and teenagers.

not be afraid to admit to problems, and seek help. A racially tense school is not conducive to learning. Hostile and punitive teachers or frightened teachers need help.

help define tasks for adolescents which provide for them a genuine chance of achieving a worthwhile goal, but not easily.

assist students, teachers, and parents to plan together what is relevant curriculum, and appropriate standards of conduct for today's world. Refuse to fight windmills of fashion, or to walk into the trap of obscenity.

be particularly sensitive to ways in which the rewards of the system are differentially distributed on the basis of race or sex, and move swiftly to eradicate inequities.

5. The education of teachers must be reoriented to take into account the significant facet of the educational problem discussed in this book. It is no longer defensible to send beginning teachers into schools with the same naiveté—to make the same mistakes that generations of teachers before them have made. The challenge of Black Power is not one to be taken up only by the social studies program with a few lessons here and there on black heroes. The total institution must be staffed by persons aware of how differences are valued, how these differences have been internalized by the individual student, and how the school can reinforce creative differences without demeaning any. This task is one for every teacher, and every person the institution employs. Reeducation of teachers can and should be shared by the school system and the college or university, and planned with individual teachers to meet their needs. Colleges, too, must be accountable for their product. And school systems must be accountable for their willingness to encourage new teachers to grow. The teacher dropout, who is often apt to be the most creative and the brightest, is a waste of social resources. Some of the highest rates of teacher loss are in the schools that need them the most. Why? If the answer is the same next year as it has been in the years past then, when students protest, riot, or burn the buildings down we have no one to blame but ourselves.

NOTES

1. J. Milton Yinger, "Recent Developments in Minority and Race Relations," *Annals of the American Academy of Political and Social Science,* 378 (July 1968), p. 135.
2. Presidents of thirty predominantly Negro colleges complained that not only had they been bypassed in fund distribution for the

education of the disadvantaged but "In fact, a major use to which these funds are put by white colleges and universities is to lure away creative black teachers and administrators from our campuses to implement their newly funded programs." *The Chronicle of Higher Education,* 3 (August 1969), p. 5; Stephen J. Wright, "Black Students and Negro Colleges: the Promise of Equality," *Saturday Review,* 51 (July 20, 1968), pp. 45 ff.

3. See, for example: John H. Bracey, August Meier, and Elliott Rudwick, eds., *Black Nationalism in America* (New York: Bobbs-Merrill, 1970); Theodore Draper, *The Rediscovery of Black Nationalism* (New York: Viking, 1969); Bob Teague, *Letters to a Black Boy* (New York: Lancer, 1969); John R. Forward and Jay R. Williams, "Internal-External Control and Black Militancy," *Journal of Social Issues,* 26 (Winter 1970), pp. 75-92.
4. See back issues of *New South* and *Southern Education Reporter.*
5. Enrollment of blacks in predominantly white public colleges and universities increased from 1.97 percent in 1968 to 2.67 percent in 1969. "Public Colleges' Negro Entrants Rise Sharply," *The Chronicle of Higher Education,* 4 (April 13, 1970), p. 1.
6. Herman D. Block, *The Circle of Discrimination: An Economic and Social Study of the Black Man in New York* (New York: New York University, 1969).
7. John H. Denton, *Apartheid, American Style* (Berkeley, Calif.: Dial, 1967).
8. "Unions Feel Growing Pressure to Take More Negroes," *U.S. News and World Report,* 56 (May 25, 1964), p. 88; U.S. Congress, House, *Equal Employment Opportunity,* Hearing before Committee on Education and Labor, 88 Congress, 1 Sess. Pursuant to H. R. Washington, 1963, p. 49.
9. Julius W. Hobson, "Uncle Sam, the Biased Employer," *Integrated Education,* 8 (September-October 1970), pp. 26-30.
10. Phyllis Myers, "The Floundering Federal Effort to Improve City Schools," *City,* 4 (June-July 1970), pp. 13-16; Washington Research Project, *Title I of ESEA: Is it Helping Poor Children?* (New York: NAACP Legal Defense and Education Fund, Inc., 1969).
11. Michael Harrington, "The Betrayal of the Poor," *Atlantic,* 225 (January 1970), pp. 71-78.

12. George A. Brager and Francis P. Purcell, eds., *Community Action Against Poverty* (New Haven: College and University, 1967).
13. See, for example: Lloyd B. Barbour, *The Black Power Revolt* (Boston: Beacon, 1968); John F. Szwed, ed., *Black America* (New York: Basic Books, 1970). See also Note 3.
14. James I. Quillen, *Textbook Improvement and International Understanding* (Washington, D.C.: American Council on Education, 1948); Ray Allen Billington and others, *The Historian's Contribution to Anglo-American Misunderstanding: Report of a Committee on National Bias in Anglo-American History Textbooks* (New York: Hobbs, Dorman, 1966).
15. Ann E. Yanko and Peter A. Kersin, (trans.), *A Soviet View of the American Past* (Madison, Wis.: State Historical Society, 1960); Donald W. Robinson, *As Others See Us: International Views of American History* (Boston: Houghton Mifflin, 1969); Donald W. Robinson, "European Textbooks and America's Racial Problem," *Social Education,* 33 (March 1969), pp. 310-313.
16. Mary R. Beard, *Woman as Force in History* (New York: Macmillan, 1946); Katherine M. Rogers, *The Troublesome Helpmate* (Seattle: University of Washington, 1966); Janice L. Trecker, "Women in U.S. History High School Textbooks," *Social Education,* 35 (March 1971), pp. 249-260.
17. Jack Abromowitz, "Textbooks and Negro History," *Social Education,* 33 (March 1969), pp. 306-309.
18. Hoyt W. Fuller, "Black Images and White Critics," *Negro Digest,* 19 (November 1969), pp. 49-50.
19. William H. Grier and Price M. Cobbs, *Black Rage* (New York: Basic Books, 1968). See also by the same authors: *The Jesus Bag* (New York: McGraw-Hill, 1971).
20. Jean D. Grambs, rev., *Black Rage,* op. cit., *Educational Product Report,* 2 (May-June, 1969), pp. 6-8; Hugh F. Butts, rev., *Black Rage,* op. cit., *Journal of Negro Education,* 38 (Spring 1969), pp. 166-168; Hugh F. Butts, rev., *Black Rage, Freedomways,* 9 (Winter 1969), pp. 59-64. For an interesting analysis of the resentment of white interpretations of black behavior, see: Alyce C. Gullattee, "The Negro Psyche: Fact, Fiction, and Fantasy," *Journal of the National Medical Association,* 61, no. 2 (1969), pp.

119-129; also Reed Whittemore, "Black Studies in Glass Houses" (review of *Amisted I), New Republic,* 162 (May 9, 1970), pp. 25-27.

21. Albert Murray, *The Omni-Americans: New Perspectives on Black Experience and American Culture* (New York: Outerbridge & Dienstfrey, 1970).
22. Hilda Taba and others, *With Focus on Human Relations; Elementary Curriculum in Intergroup Relations; Curriculum in Intergroup Relations: Secondary School; Reading Ladders for Human Relations; Sociometry in Group Relations; Diagnosing Human Relations Needs; Literature for Human Understanding; With Perspective on Human Relations; Leadership Training in Intergroup Education; School Culture; Intergroup Education in Public Schools* (Washington, D.C.: American Council on Education, 1947-1955).
23. Allison Davis, Burleigh B. Gardner, and Mary R. Gardner, *Children of Bondage,* 1940; E. F. Frazier, *Negro Youth at the Crossways,* 1940; C. S. Johnson, *Growing Up in the Black Belt,* 1941; I. DeA. Reid, *In a Minor Key,* 1940; R. L. Sutherland, *Color, Class and Personality,* 1942; W. L. Warner, B. H. Junker, and W. A. Adams, *Color and Human Nature,* 1941 (Washington, D. C. : The American Council on Education).
24. Helen G. Trager and Marian Radke Yarrow, *They Learn What They Live* (New York: Harper & Brothers, 1952).
25. W. Lloyd Warner, Robert J. Havighurst, and Martin B. Loeb, *Who Shall Be Educated?* (New York: Harper & Brothers, 1944).
26. Alice V. Keliher, *A Critical Study of Homogeneous Grouping* (New York: Columbia, Teachers College, 1931).
27. Leonard P. Ayers, *Laggards in Our Schools: A Study of Retardation and Elimination in City School Systems* (New York: Russell Sage Foundation, 1909).
28. See, for example: Davis McEntire, *The Population of California* (San Francisco: The Commonwealth Club of California, 1946).
29. C. Vann Woodward, *The Strange Career of Jim Crow* (New York: Oxford, 1955).
30. Dixon Wecter, "Commissars of Loyalty," *Saturday Review of Literature,* 33 (May 13, 1950); George R. Stewart, *The Year of the Oath: The Fight for Academic Freedom at the University of California* (New York: Doubleday, 1950); Howard K. Beale, *Are*

American Teachers Free? (New York: Scribner, 1936); Vivian T. Thayer, *The Attack upon the American Secular School* (Boston: Beacon, 1951).

31. Herbert Aptheker, *American Negro Slave Revolts* (New York: International Publishers, (1943, 1963).
32. Carl L. Becker, *The Declaration of Independence* (New York: Alfred A. Knopf, 1942), pp. 166-167.
33. See: bibliography in Jean D. Grambs, *Intergroup Education: Methods and Materials* (Englewood Cliffs, N.J.: Prentice-Hall, 1968); also Bernard Goldstein, *Low Income Youth in Urban Areas: A Critical Review of the Literature* (New York: Holt, 1967); Elizabeth W. Miller, *The Negro in America: A Bibliography* (Cambridge, Mass.: Harvard, rev. ed., 1970); Meyer Weinberg, *Desegregation Research: An Appraisal* (Chicago: Integrated Education Associates, 2nd rev. ed., 1970); Melvin Tumin, *Research Annual on Intergroup Relations* (New York: Anti-Defamation League of B'nai B'rith, 1970).
34. Jean D. Grambs, "Negro Self-Concept," in William Kvaraceus et al., *Negro Self-Concept* (New York: McGraw-Hill, 1965).
35. G. Louis Heath, "An Inquiry into a University's 'Noble Savage' Program," *Integrated Education,* 8 (July-August 1970), pp. 4-9.
36. Hertha Riese, *Heal the Hurt Child* (Chicago: University of Chicago, 1962).
37. Seymour Sarason et al., *Psychology in Community Settings* (New York: Wiley, 1966), p. 281.
38. Nancy Hicks, "The Toll of Hunger on a Child's Intelligence," *New York Times,* March 1, 1970, p. 13E; Rita Bakan, "Malnutrition and Learning," *Phi Delta Kappan,* 51, no. 10 (June 1970), pp. 529-530.
39. Hicks, op. cit.
40. Mary Frances Greene and Orletta Ryan, *The Schoolchildren* (New York: Pantheon, 1965), p. 44.
41. *The New York Times,* July 14, 1968.
42. The *Washington Post,* June 3, 1970; A bill outlawing use (i.e., where children might get at it) of such paint was signed into law in Maryland, May 17, 1971.
43. Eleanor B. Leacock, *Teaching and Learning in City Schools* (New York: Basic Books, 1969), p. 79.
44. Robert Rosenthal and Lenore F. Jacobson, *Pygmalion in the*

Classroom: Self-Fulfilling Prophecies and Teacher Expectations (New York: Holt, 1968); Albert H. Yee, *Journal of Educational Psychology,* 49, no. 4 (1968).

45. Edmund V. Gordon, "Introduction," *Review of Educational Research,* 40 (February 1970), pp. 1-12.
46. Alexander Moore, "The Inner-City High School, Instructional and Community Roles," in Harry Passow, ed., *Reaching the Disadvantaged Learner* (New York: Teachers College, 1970), p. 225.
47. David Gottlieb, "Teaching and Students: The Views of White and Negro Teachers," *Sociology of Education,* 37 (Summer 1964), pp. 345-353. See also: David E. Hunt and John Dopyera, "Personality Variations in Lower Class Children," *Journal of Psychology,* 62 (1966), pp. 47-54.
48. Moore, op. cit., p. 225.
49. Leacock, op. cit., pp. 137, 195-196.
50. Ibid, p. 203; for another research study which supports these findings see: Jerome Beker, James B. Victor, and Linda F. Seidel, "School Days," in Irwin Deutscher and Elizabeth Thompson, eds., *Among the People* (New York: Basic Books, 1968), pp. 174-208. See also: Margaret E. Hertzig et al., "Class and Ethnic Differences in the Responsiveness of Preschool Children to Cognitive Demands," *Monographs of the Society for Research in Child Development,* 33, no. 117 (1968), pp. 1-69. This study of middle-class and Puerto Rican preschoolers suggests that linguistic patterns rooted in cultural differences may be perceived as IQ differences when this may be open to serious question.
51. Abraham Tannenbaum, *Adolescent Attitudes Toward Academic Brilliance* (New York: Teachers College, 1962).
52. Thomas F. Pettigrew, "The Negro and Education: Problems and Proposals," in Irwin Katz and Patricia Gurin, eds., *Race and the Social Sciences* (New York: Basic Books, 1969), p. 77.
53. Irwin Katz, "Experiments in Negro Performance in Bi-Racial Settings," in Matthew Miles and W. W. Charters, Jr., eds., *Learning in Social Settings* (Boston: Allyn and Bacon, 1970), pp. 225-239.
54. Rosenthal and Jacobson, op. cit.
55. Miriam Goldberg, A. Harry Passow, and Joseph Justman, *The Effects of Ability Grouping* (New York: Teachers College, 1966),

p. 165; Wallace La Benne and Bert I. Greene, *Educational Implications of Self-Concept Theory* (Pacific Palisades, Calif.: Goodyear, 1969), pp. 48-64. In a carefully devised study, Chang and Raths found that one of the key factors resulting in measurably lower achievement for 'lower track' children remaining in school was the fact that teachers of less able students taught and concentrated on the same school tasks as those teaching able learners. The learning deficits of the less able were therefore ignored, not remedied. The result was that the longer children stayed in school the less able they were to learn. Sunnyuh Shin Chang and James D. Raths, "The Schools' Contribution to the Cumulating Deficit," *Journal of Educational Research,* 64 (February 1971), pp. 272-273.

56. The *Washington Post,* September 27, 1970; September 28, 1970; October 1, 1970; October 6, 1970 (Discussions of the Clark plan).
57. Raymond L. Jerrems, "Racism: Vector of Ghetto Education," *Integrated Education*, 8 (July-August 1970), pp. 40-47.
58. James M. Coleman et al., *Equality of Educational Opportunity* (Washington, D.C.: Government Printing Office, 1966); Robert A. Dentler, "Equality of Educational Opportunity: A Special Review," *The Urban Review,* 1, no. 5 (1966), pp. 27-29.
59. M. Brewster Smith, "Explorations in Competence: A Study of Peace Corps Teachers in Ghana," *American Psychologist,* 21, no. 6 (1966), pp. 555-566. See also by the same author: "Competence and Socialization," in John A. Clausen, ed., *Socialization and Society* (Boston: Little, Brown, 1968), pp. 272-320.
60. Erik Erikson, *Identity, Youth and Crisis* (New York: Norton, 1968), pp. 107-114.
61. Robert W. White, "Competence as a Basic Concept in the Growth of Personality," unpublished paper prepared for the Social Science Research Council Conference on the Socialization of Competence, 1965.
62. Beker et al., op. cit., p. 205.
63. Abraham H. Maslow, *Toward a Psychology of Being* (New York: Van Nostrand, 1962).
64. John W. Gardner, *Self-Renewal: The Individual and the Innovative Society* (New York: Harper & Row, 1964).
65. Charles W. Thomas, "Boys No More: Some Social Psychological

Aspects of the New Black Ethic," *The American Behavioral Scientist,* 12 (March-April 1969), pp. 38-42.

66. David Birch and Joseph Veroff, *Motivation: A Study of Action* (Belmont, Calif.: Brooks/Cole, 1966), p. 63.
67. Nathan Glazer and Daniel P. Moynihan, *Beyond the Melting Pot: the Negroes, Puerto Ricans, Jews, Italians, and Irish of New York City,* 2nd ed., (Cambridge: M. I. T., 1970).
68. See, for example: Carey Winfrey, "Volume I, Number 1—Birth of a New Black Magazine," *New York,* 3 (April 27, 1970), pp. 58-60. This article describes *Essence,* a new magazine for the "Young, urban, inquisitive, and acquisitive black woman."
69. "Student Strikes: 1968-69," *Black Scholar,* 1 (January-February 1970), pp. 65-75.
70. J. Herman Blake, "Black Nationalism," *Annals of the American Academy of Political and Social Science,* 382 (March 1969). See also: Note 76.
71. Commission on Professional Rights and Responsibilities, *Wilcox County, Alabama: A Study of Social, Economic, and Educational Bankruptcy,* (Washington, D.C.: National Education Association, June 1967).
72. Nathan Hare, "Black Students and Negro Colleges: The Legacy of Paternalism," *Saturday Review,* 51 (July 20, 1968), pp. 44 ff; Orlando L. Taylor, "New Directions for American Education: a Black Perspective," *The Speech Teacher,* 19 (March 1970), pp. 111-116; Alvin A. Poussaint, "Why Blacks Kill Blacks," *Ebony,* 25 (October 1970), pp. 143-150.
73. *Time* Magazine, January 12, 1970, p. 59.
74. Bayard Rustin, "The Failure of Black Separatism," *Harper's Magazine,* 240 (January 1970), pp. 25-29.
75. "Most College Grads Still Want Jobs 'Where the Money Is' " *Ebony,* 25 (May 1970), pp. 42-51.
76. Nathan Caplan, "The New Ghetto Man: A Review of Empirical Studies," *Journal of Social Issues,* 26 (Winter 1970), pp. 65-66.
77. Eldridge Cleaver, *Soul on Ice* (New York: McGraw-Hill, 1968); Frantz Fanon, *Black Skin, White Mask* (New York: Grove, 1967). For an assessment of Fanon's writing and influence see: Dennis Forsythe, "Frantz Fanon: Black Theoretician," *Black Scholar,* 1 (March 1970), pp. 2-10. See also: Robert L. Scott and Wayne

Brockridge, *The Rhetoric of Black Power* (New York: Harper & Row, 1969).

78. Francine du Plessix Gray, "The Panthers at Yale," *New York Review of Books,* 14 (June 4, 1970), pp. 29-36; Leon Friedman, *Southern Justice* (New York: Pantheon, 1965); Jacobus ten Broek and the editors of the California Law Review, eds., *The Law of the Poor* (San Francisco: Chandler, 1966); Truman Nelson, *The Torture of Mothers* (Boston: Beacon, 1965).
79. Jack Nelson and Jack Bass, *The Orangeburg Massacre* (New York: World, 1971). The shootings at Jackson State College, Jackson, Mississippi, are still a source of local black anger: Graig Vetter, "Funeral in Jackson," *Playboy,* 18 (June 1971), pp. 149-152; *Washington Post,* May 16, 1971.
80. Arthur E. Hippler, "The Game of Black and White at Hunters Point," *Transaction,* 7 (April 1970), pp. 56-63. See also: Richard Hall, "Dilemma of the Black Cop," *Life,* 69 (September 18, 1970), pp. 60-69.
81. Mitchell Gordon, *Sick Cities* (Baltimore: Penguin, 1965); Jeanne R. Lowe, *Cities in a Race with Time* (New York: Vintage, 1968); Ulf Hannerz, *Soulside: Inquiries in Ghetto Culture and Community* (New York: Columbia, 1969).
82. Dr. Frank Pumphrey, State Board of Education, Maryland, statement, November 1970.
83. See: Jean Dresden Grambs and John C. Carr, eds., *Black Image: Education Copes with Color* (Dubuque, Iowa: Wm. C. Brown, 1971), for a list of black journals.
84. Ibid., Appendix II.
85. See, for example: Calvin Trillin, "U.S. Journal: Dorchester County, S. C., Victoria Delee—in Her Own Words," *New Yorker,* 47 (March 27, 1971), pp. 86-92.
86. E. Earl Baughman, *Black Americans* (New York: Academic, 1971), p. 41.
87. Ibid., p. 44.
88. Wilbur B. Brookover and Edsel L. Erickson, *Society, Schools and Learning* (Boston: Allyn and Bacon, 1969), p. 106.
89. Edward M. Glaser and Harvey L. Ross, *A Study of Successful Persons from Seriously Disadvantaged Backgrounds: Final Report,* prepared for the Office of Special Manpower Programs,

Washington, D.C., Contract No. 82-05-68-03. (Washington, D.C.: Department of Labor, March 31, 1970), p. 75.

90. Op. cit., p. 75.
91. Robert E. Staples, "Educating the Black Male at Various Class Levels for Marital Roles," *The Family Coordinator,* 19 (April 1970), pp. 164-167.
92. Joan C. Baratz and Roger W. Shuy, *Teaching Black Children to Read* (Washington, D.C.: Center for Applied Linguistics, 1969).
93. E. Franklin Frazier, *Negro Youth at the Crossways* (Washington, D.C.: American Council on Education, 1940).
94. J. Richard Udry, Karl E. Bauman, and Charles Chase, "Skin Color, Status, and Mate Selection," *American Journal of Sociology,* 76 (January 1971), pp. 722-732.
95. H. Edward Ransford, "Skin Color, Life Chances, and Anti-White Attitudes," *Social Problems,* 18 (Fall 1970), pp. 164-178.
96. Ibid., p. 178.
97. Ann F. Brunswick, "What Generation Gap?" *Social Problems,* 17 (Winter 1970), pp. 358-371.
98. Ibid., p. 369.
99. Erving Goffman, *Stigma: Notes in the Management of Spoiled Identity* (Englewood Cliffs, N.J.: Prentice-Hall, 1963).
100. Warner, Havighurst, and Loeb, op. cit.; Patricia Sexton, *Education and Income* (New York: Viking, 1961).
101. Robert J. Havighurst and Hilda Taba, *Adolescent Character and Personality* (New York: Wiley, 1949). See also: Robert F. Peck and Robert Havighurst, *The Psychology of Character Development* (New York: Wiley, 1960).
102. Richard W. Brozovich, "Characteristics Associated with Popularity Among Different Racial and Socioeconomic Groups of Children," *Journal of Educational Research*, 63 (July-August 1970), pp. 441-444; Monroe K. Rowland and Phillip DelCamp, "The Values of the Educationally Disadvantaged: How Different are They?" *Journal of Negro Education,* 37 (Winter 1968), pp. 86-89; Lee Rainwater, "The Problem of Lower Class Culture," *Journal of Social Issues,* 26, no. 2 (1970), pp. 133-148.
103. Harry I. Miller and Roger R. Woock, *Social Foundations of Urban Education* (Hinstead, Ill.: Dryden, 1970), pp. 398-400. See also: Anthony T. Soares and Louise M. Soares, "Self-Perceptions of

Culturally Disadvantaged Children," *American Educational Research Journal*, 6 (January 1969), pp. 31-45.

104. Goldberg, Passow, and Justman, op. cit.
105. The simple designations of third grade, tenth grade, are equally misleading. These classifications tend to reduce the differences; in fact, as noted by Leacock, op. cit., they serve to deny and punish differences particularly at the upper end of the ability continuum.
106. Leacock, op. cit.
107. Louis M. Smith and William Geoffery, *Complexities of the Urban Classroom* (New York: Holt, 1969).
108. Elizabeth Eddy, *Walk the White Line: A Profile of Urban Education* (Garden City, N.Y.: Anchor, 1967).
109. Estelle Fuchs, *Teachers Talk: Views from Inside City Schools* (New York: Anchor, 1967).
110. James Herndon, *The Way It Spozed To Be* (New York: Simon and Schuster, 1968); John Holt, *The Underachieving School* (New York: Pitman, 1969); Herbert R. Kohl, *The Open Classroom* (New York: New York Review/Vintage Books, 1969); Herbert R. Kohl, *Thirty-Six Children* (New York: New American Library/World, 1967); Jonathan Kozol, *Death at an Early Age: The Destruction of the Hearts and Minds of Negro Children in the Boston Public Schools* (Boston: Houghton Mifflin, 1967); Sunny Decker, *An Empty Spoon* (New York: Harper & Row, 1969).
111. George Spindler, *The Transmission of American Culture* (Cambridge, Mass.: Harvard, 1959).
112. Gottlieb, op. cit.; Helen M. Davidson and Gerhard Long, "Children's Perceptions of Their Teachers' Feelings Toward Them Related to Self-Perception, School Achievement and Behavior," *Journal of Experimental Education,* 29 (December 1960), pp. 107-118.
113. Jean D. Grambs and Walter B. Waetjen, "The Right to be Equally Different: a New Right for Boys and Girls," *National Elementary Principal,* 53 (November 1964).
114. Paul H. Bowman, "Improving the Pupil Self-Concept," in Robert D. Strom, ed., *The Inner-City Classroom: Teacher Behaviors.* (Columbus, Ohio: Charles E. Merrill, 1966). See also: Walter E. Schafer, "Participation in Interscholastic Athletics and Delinquency," *Social Problems,* 17 (Summer 1969), pp. 41-47.

115. Leacock, op. cit., pp. 190-191.
116. Dante P. Ciochetti, "CBE Interviews: Kenneth B. Clark," *Bulletin of the Council for Basic Education,* 14 (November 1969), pp. 15-16.
117. M. Brewster Smith, "Normality for an Abnormal Age" (in press).
118. Helen E. Rees, *Deprivation and Compensatory Education* (Boston: Houghton Mifflin, 1968).

Name Index

Subject Index